Textbook of

ADVANCED PHACOEMULSIFICATION TECHNIQUES

Edited by

Paul S. Koch, MD

James A. Davison, MD

SLACK International Book Distributors

In Japan
  Igaku-Shoin, Ltd.
  Tokyo International P.O. Box 5063
  1-28-36 Hongo, Bunkyo-Ku
  Tokyo 113
  Japan

In Canada
  McGraw-Hill Ryerson Limited
  300 Water Street
  Whitby, Ontario
  L1N 9B6
  Canada

In all other regions throughout the world, SLACK professional reference books are available through offices and affiliates of McGraw-Hill, Inc. For the name and address of the office serving your area, please correspond to

  McGraw-Hill, Inc.
  Medical Publishing Group
  Attn: International Marketing Director
  1221 Avenue of the Americas —28th Floor
  New York, NY 10020
  (212)-512-3955 (phone)
  (212)-512-4717 (fax)

Editorial Director: Cheryl D. Willoughby
Publisher: Harry C. Benson

Printed in the United States of America

Library of Congress Catalog Card Number: 89-43329

ISBN: 1-55642-158-3

Published by: SLACK Incorporated
            6900 Grove Road
            Thorofare, NJ 08086-9447

Last digit is print number: 10 9 8 7 6 5 4 3 2

To our wives, Joanne and Vicki, and to our children Paul, Katie, Adam, Megan and Lindsey

—Paul S. Koch, MD
—James A. Davison, MD

Table of Contents

Contributors

Alan S. Crandall, MD

James A. Davison, MD

David M. Dillman, MD

I. Howard Fine, MD

William J. Fishkind, MD

Robert M. Kershner, MD

Paul S. Koch, MD

Robert P. Lehmann, MD

William F. Maloney, MD

Marc A. Michelson, MD

Jack A. Singer, MD

Alan W. Solway, MD

Preface

The Story of Paul & Jim & Steve & George
by Paul S. Koch, M.D.

Do you remember the story about how Steven Spielberg and George Lucas teamed up to make the movie Raiders Of The Lost Ark? Neither do I, exactly, but they met at some resort or another and sat around the swimming pool, or maybe it was on the beach. Wherever it was, they were drinking gin and tonics, or iced tea, or maybe they were just thirsty, and they began talking about the great serials they used to watch at the movies, and pretty soon they decided to make one of their own, only it would be just one movie, not a lot of short ones, but it turned out to be three of them.

Anyway, I was sitting on the Cote d'Azur sipping a Coke or a Pepsi or whatever they had when my buddy Jim Davison came along and we started talking about phacoemulsification, because that's our line of work—like movies are to Steve and George. I allowed as to how I had been pursuaded, in a weak moment, to follow up my basic textbook of phacoemulsification with a more advanced one, for those who had followed my precepts and were ready for the more and better.

Slyly, I told Jim how I was a big fan of his writings and how no book like this could possibly be complete without his contributing some modest morsel. He leaped or leapt to his feet and exclaimed that it would be his privilege. I leaped or leapt to mine and said that, if that was the case, he would be even more privileged if he contributed a lot; and he said "right" and I said "right," and pretty soon we were working on this book together.

And that's how maybe we became kind of like the Spielberg and Lucas of cataract surgery, and I'd be pretty sure one way or another if I could remember if Steve and George got together the way we did.

The End

Introduction

It was a pleasure putting this book together with Jim Davison. If you know him, I think you'll agree that he has one of the finest minds in cataract surgery and that he is a great surgeon to boot. My respect for him and his influence on me are both considerable.

If you took the time to read the little story that makes up the preface to this book, you should know that it's not exactly the truth, but my daughter, Katie (age 7), thought it was a cute story, so we left it in. What is true, though, is that I urged Jim to get involved more than he ever imagined.

We flipped a coin. He won. I get to write the introduction to this book, but he gets to write the introduction to the section on Endocapsular Phacoemulsification.

About ten years ago I was making the difficult conversion from intracapsular surgery directly to phacoemulsification. I wasn't doing too badly, considering that I had to learn a lot about the anatomy of the lens. Until that time I had no idea that there was such a thing as cortex, or how big a nucleus was, or how much I could handle (manhandle? mishandle?) the capsule.

In 1980 and 1981, I don't think you could pick up a copy of *Ophthalmology Times* without seeing some sort of article by Dr. Robert Welsh talking about this amazing guy he discovered, a Dr. James Gills, who was doing manual extracapsular surgery with a syringe and a cannula, in minutes, for pennies, with guaranteed 100% in–the–bag lens placement.

Eventually, I had to face the fact that there was an awful lot about extracapsular surgery that I didn't understand. In May 1981, I took the Gills/Welsh course in New Port Richey, Florida.

I was totally blown away. I never realized that there was so much to know about the anatomy of the lens, how the nucleus could be separated into cleavage planes, how the cortex could be stripped and teased. All of the things I had been doing was because I copied someone. For the first time I understood what I was supposed to do and what I was doing it to. Incredible!

The 1981 Gills/Welsh course was the single most remarkable couple of days I have had in my ophthalmology career. It completely changed the way I looked at a human lens.

Years passed. I did a lot of phacoemulsification, taught some courses, wrote a book. My friend, Dr. Bill Maloney, called me on the phone. Would I, he asked, like to help him set up a new course, one he guaranteed would be special. I would have to fly to California (I hate flying) and spend a couple of days sequestered with the faculty in a studio where we would do some video-taping for a teaching tape.

I went to California and met with Bill and the rest of his group, Drs. Dave Dillman, Marc Michelson, and William Fishkind. Bill sat us down on the first day and ordered us, no, urged us, no wait, encouraged us to reach a consensus on what was the single best way for a beginner to perform each step of the operation. What was the best incision, the best capsulotomy, the best prolapse, and so forth? Each step had to be set up so that a beginner could abort the case any time it passed a certain comfort level.

From this session, Bill set up the *Three Steps To Phaco* series of courses.

From the start, he set up some rules. The ability to perform phacoemulsification had to be developed through a learned process. We could never consider it to be some sort of a special procedure capable of being performed only by those chosen through some special birthright. There are no preordained supersurgeons who would become phaco surgeons. Anybody could.

The courses were designed to teach phacoemulsification, not just to demonstrate it. We would teach one technique, exactly standardized from faculty member to faculty member and consistent from course to course. No matter what each of us were doing in our own practices, at the courses there would be only one "best" way to learn phacoemulsification.

Then something special happened. This faculty, plus the others who joined us—Drs. Alan Crandall, Robert Kershner, Alan Solway, Robert Lehmann, Richard Livernois, and Roderick Morgan—turned into an ensemble, a touring company, a road show. Routines developed, jokes, stories, gag videos, fun.

Imagine, fun at a medical meeting.

Bill's *Three Steps To Phaco* courses turned into the very best instructional meetings of any kind. Period. I'm especially proud because I was fortunate to be dragged along behind him, but Bill's courses were very, very special, even magical.

By the end of 1988, like a lot of other people, I was playing with endocapsular surgery. Marc Michelson was way ahead of me. In December 1987, we sat next to each other in Los Angeles when Dr. Jay Patel gave one of his early talks on intercapsular surgery. I thought it was a lot of work. Marc thought it was the future. I went home and did my thing. Marc went home and did a new thing.

Then, in April 1989, Bill had a course in New York City. For the first time, Dr. Howard Fine and I were on the same panel. Howard showed a video of his "Chip and Flip" technique, and I went wild. I went nuts. I felt just like I did in New Port Richey eight years before. Howard defined for me an entirely new strategy for looking at a nucleus.

Howard showed me that the lens was like an onion and that it was possible to separate the inner layers from the outer layers. He could dissect out the central hard nuclear core, emulsify that while working within a fairly thick cushion of outer nucleus, and then pull the outer nucleus out later. This gave him a level of control and safety magnitudes above the techniques I was working on. It changed, once again, the way I looked at a human lens. By comparison, it made the stuff I'd been doing and teaching for years, the iris plane techniques, the prolapse and rotation techniques, even some of the endocapsular methods I was working on, look barbaric.

I jumped into Chip & Flip and right away noticed that the operation was easier, my pulse slower, and the results better than anything I had ever seen before. My personal technique, "Spring Surgery," is a variation of and indebted to Howard's technique.

Remember one thing as you delve into this book. These are advanced techniques, at least by 1990 standards. You should not adopt these techniques if you are learning phacoemulsification. Unlike the Three Steps method, there are not a lot of points where you can bail out if you need to. These techniques are just not as safe for a beginnner. Sorry.

If you're new to this and you want to do endocapsular phacoemulsification, learn the iris plane stuff first and then switch over. You will be able to bail out easier if you need to than you could with the endocapsular procedures. Let's face it, if you're bummed out enough to want to bail out, you're bummed out enough to want to bail out as easily as possible. Why complicate your life by locking yourself into an endocapsular technique before you're ready?

One more thing. Each of the chapter authors is an excellent teacher in his own right. Each of us spends a lot of time out there on the teaching circuit. We all agree that this book should teach advanced phacoemulsification techniques, not just demonstrate them. If you want to learn something and we don't get the point across in a way that makes you understand it, write to any of us and we will clarify it. That's a promise.

And now the book. It's a collection of essays, themes, and chapters covering a variety of topics in what is considered the field of advanced phacoemulsification techniques. In some cases the authors have been revising their contributions right up to the publication date to be sure the information is as current as possible. This is a rapidly developing field; I hope you will be caught up in its energy. Good luck.

Paul S. Koch, M.D.

Acknowledgment

Special thanks to Kayla Danielson, RN and Sherri Tolatovicz for photographic and editorial assistance.

Section I
PREPARATION

1

Setup

James A. Davison, MD

A high quality microscope with a footswitch-controlled X-Y adjustment platform and focus-zoom control is essential. A total power of 7x to 15x is appropiate for capsulorhexis and nucleus dissection in capsular bag phacoemulsification. Many times the highest magnification possible will be needed for capsulorhexis. A lower magnification is usually desirable for much of the phacoemulsification so that the surgeon can have an awareness of the "big picture" of what is going on in the posterior chamber as well as the rest of the field. Most surgeons will select a 175–mm objective lens for use during surgery. Most use traditionally angled optics as well so that they are looking down slightly at the operative field.

In order to perform well, this must be individualized. The surgeon must be comfortable. I am six feet, two inches tall and have a long back, so I prefer quite a different microscope arrangement. I use a 250–mm objective lens and a variable angle ocular configuration. I even mount the ocular accessory upside down so that I can sit tall and comfortably without having to flex my neck or back (Figure 1–1). When I am operating, I am sitting tall with my arms at my side and I am looking straight ahead without hunching over (Figure 1–2). It is important for me to be comfortable. The upright position is important. I can use my arms and hands comfortably, and my legs and feet are free to move and operate the foot controls because they are not needed to supply balance or support my upper body.

A high quality, reliable phacoemulsification machine with footswitch linear control phacoemulsification and suction is necessary to perform this technique successfully. The 45-degree ultrasonic tip is ideal for isolating and facilitating the independent functions of cutting, suction, and manipulation. Also, to best accomplish these functions safely, I like to change two of the machine's standard factory preset settings. When using the Coopervision Ten Thousand Series machines, I increase the vacuum to 61 mm Hg during phacoemulsification. I reduce the flow rate, however, to 17 cc/min. vs. the preset of 25 cc/min. (Letter to the editor. Personal correction of a phacoemulsification machine problem.

J Cataract Refract Surg 14: 456–458, 1988). The reduced flow rate has several advantages.

The operation proceeds a little slower in a more controlled fashion. There is a reduced likelihood of capsular aspiration at the endstages of phacoemulsification. Diaphragmatic movements of the iris and posterior capsule are not as obvious with this reduced flow rate.

The increased vacuum is adequate to suck any material into the tip. The tip can become occluded by any firm flat nuclear plate when presented to the aperture. Brief taps of low power emulsification energy will eat into the plate without having to increase the vacuum setting.

There is much variation, as most surgeons who practice capsular bag phacoemulsification use a 30–degree tip and the factory preset setting. Tennis shoes or some other footwear that affords a sense of touch should be worn to facilitate an awareness of pedal position and the fine movement required for this technique. Stiff soled street shoes are not recommended. I like to use my right foot for the microscope control, which I feel is a more complex function than the phacoemulsification footswitch control (Figures 1–3, 1–4). I usually make about fifty adjustments on the microscope pedal during a case.

The surgical suite should be arranged so that everyone on the surgical team can be easily aware of the others' activity (Figure 1–5). The surgeon sitting at the head of the table can communicate with everyone in the room. He is

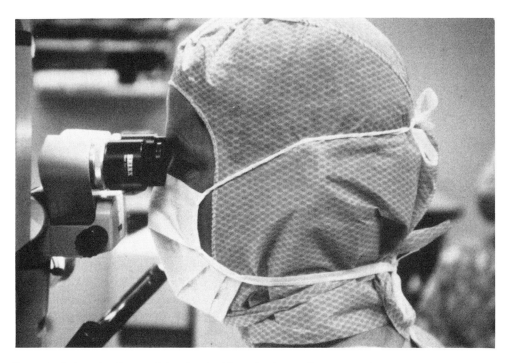

Figure 1–1. I am looking straight ahead without having to flex my cervical spine. The microscope oculars are mounted upside–down so I can get a little extra height.

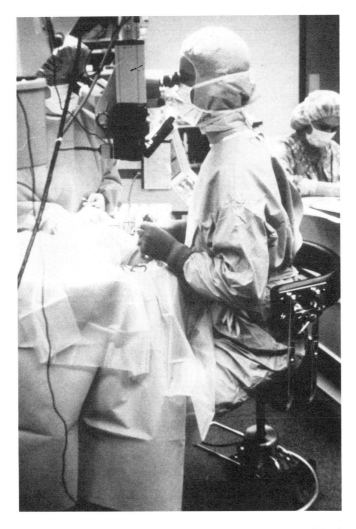

Figure 1–2. The extra working distance pays off for me because I can sit tall in the saddle without hunching. Crouching over the surgical field eventually fatigues the surgeon.

aware of every sound coming from the patient, the phacoemulsification machine, and the surrounding support personnel. He is also monitoring much of this with his peripheral vision. The anesthesiologist or nurse anesthetist sits at the patient's left so that all of his machinery is isolated and independent from the surgical equipment. The scrub nurse usually stands so that he or she has access to the phacoemulsification machine tray, Mayo stand, and back table. The circulating nurse can perform the important machine programming functions while doing minor paperwork and being available to bring new items into the surgical field. A video monitor not only adds to the pleasure of the surgical experience but, by showing the surgery, facilitates a necessary awareness by the staff of the surgical situation (Figure 1–6).

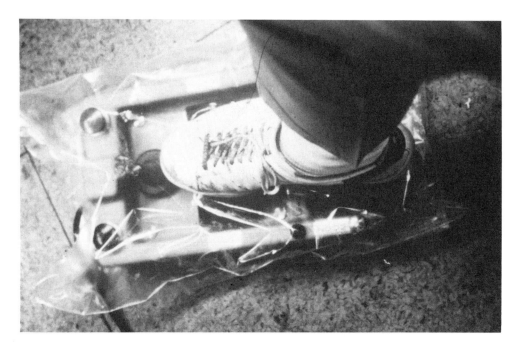

Figure 1–3. I am right–handed and right–footed. The more complex function of adjusting the microscope is delegated to the more capable right foot. Tennis shoes allow a sense of "touch" that cannot be obtained otherwise. It is very important to see perfectly throughout the operation. This can only be accomplished with the correct field composition, focus, and magnification.

Figure 1–4. My left foot runs the multiple–function linear–control phacoemulsification pedal. The perfect functioning of this pedal is critical to the success of the procedure, but its accomplishment is actually less complex than the microscope function.

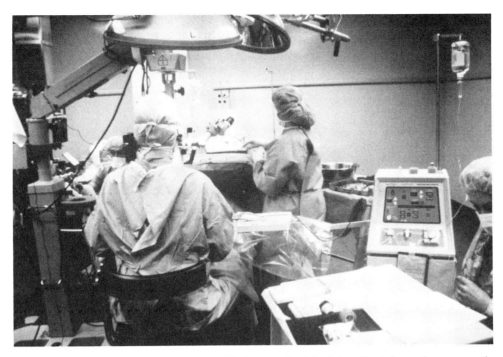

Figure 1–5. The operating room is set up with the anesthesiologist on the left, scrub nurse on the right, and the circulating nurse on the far right, with the surgeon at the head of the table.

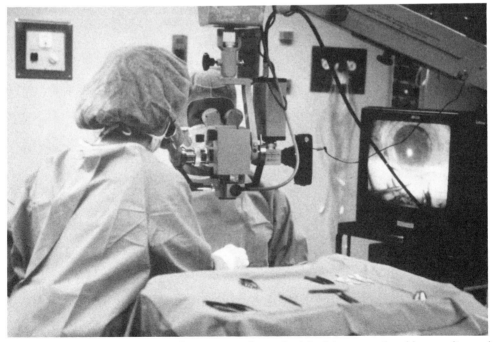

Figure 1–6. The scrub nurse, circulating nurse, and anesthesiologist can see the video monitor and appreciate what's going on during surgery.

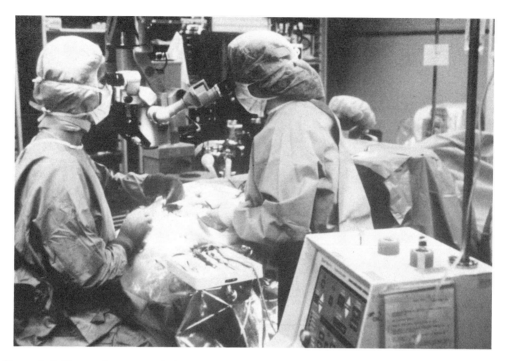

Figure 1–7. The scrub nurse can help during the operation by looking through the microscope. Instruments are available on the phaco tray and the Mayo stand. The surgeon can look over and see the console of the phaco machine.

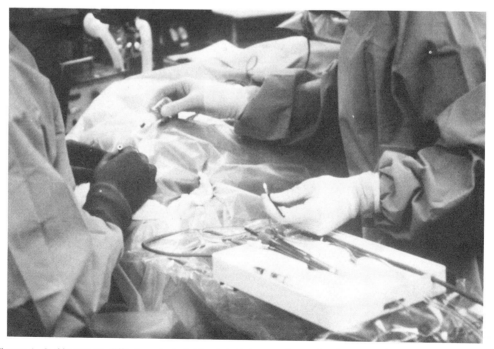

Figure 1–8. Almost no difference in height exists between the patient's eye and the instrument tray.

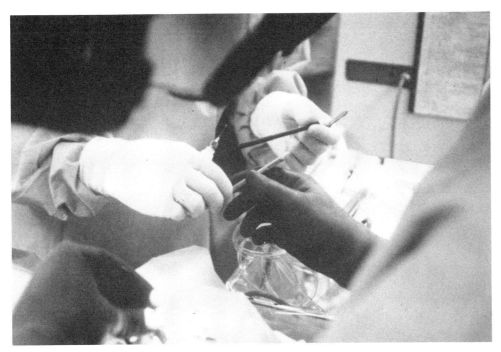

Figure 1–9. The surgeon's arm can extend to the right to receive new instruments without changing height. The action resembles that of a tone arm on a record turntable.

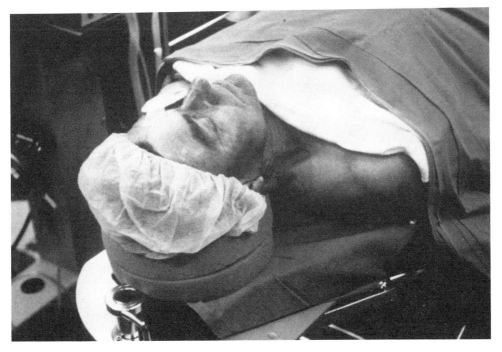

Figure 1–10. The Chan wrist rest is in place. Two donut foam pillows secure the patient's head at the proper height.

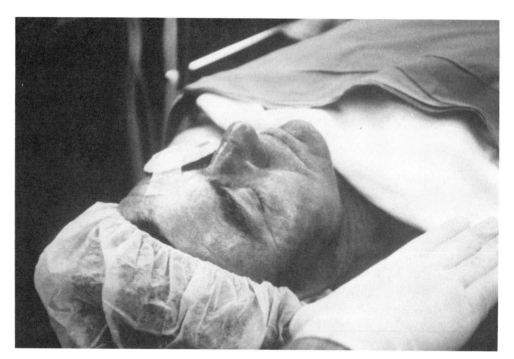

Figure 1–11. The patient's face is parallel with the ceiling. This makes him comfortable and allows comfortable access with the phacoemulsification handpiece.

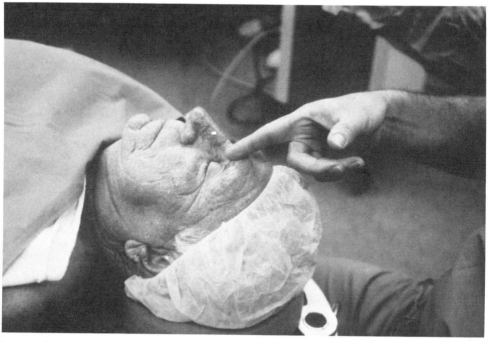

Figure 1–12. Steep brows may require a slight neck extension to allow good phaco tip access. Do not overcompensate. It is easier for a patient to extend his or her neck rather than flex it for you in a positioning effort during surgery.

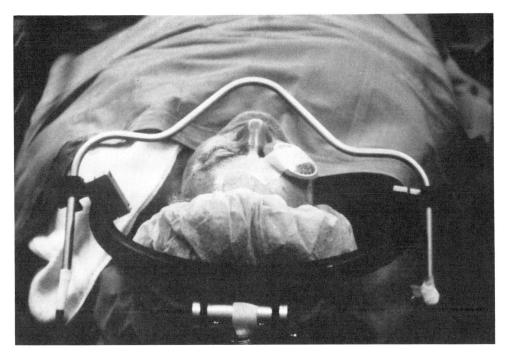

Figure 1–13. The wrist rest (water trough) is in place with the perforated breathing bar above. A Fox shield has been placed over the eye not to be operated and the head slightly turned toward the nonoperated eye.

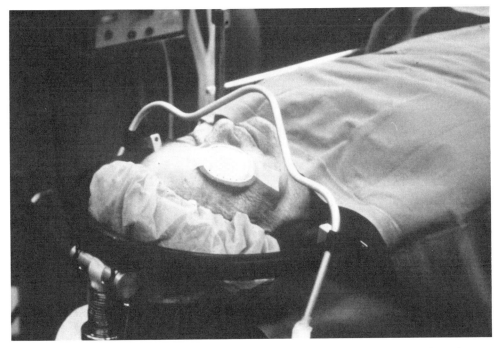

Figure 1–14. The breathing bar is angled slightly downward toward the feet and away from the nose to give the scrub nurse and surgeon more room.

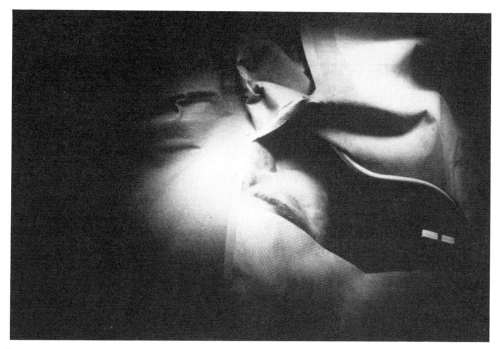

Figure 1–15. A paper drape is being applied. Note that the vertical midline of the drape is applied to the center of the bridge of the nose. The superior drape is above the brow.

Figure 1–16. After the horizontal slit is made in the drape, a vertical snip is cut in the center, one above and one below.

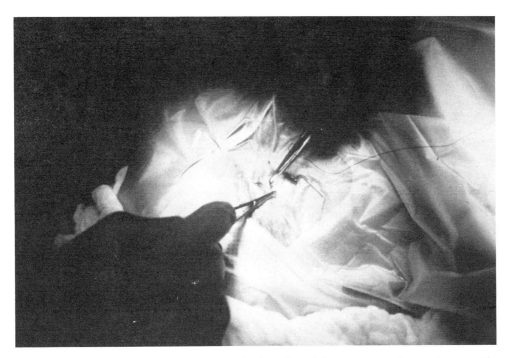

Figure 1–17. Kirby forceps are used to grasp the insertion of the rectus muscles, and non–cutting needle 4–0 silk sutures are placed in each.

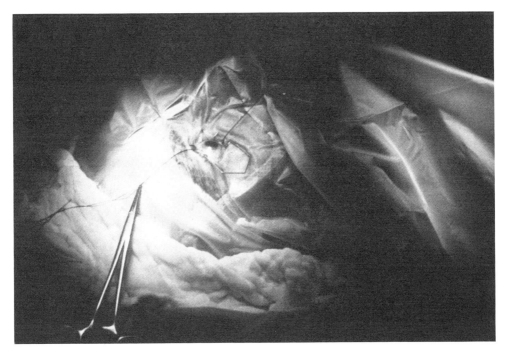

Figure 1–18. The eye is prepared for surgery.

The cart must be at the correct height so that the eye is at the appropriate level on the phacoemulsification machine, as determined by the operator's manual. All of the machine's programmed parameters must be "true". The bottle height, suction, etc., are all calibrated based on the eye being at a certain level with respect to a level referenced on the machine. Most are designed so that the eye should be at the same height as the aspiration pump. The accessory instrument tray is usually at this height so that instruments can be passed horizontally with no vertical movement of the surgeon's or assistant's arm (Figures 1–7, 1–8, 1–9).

The patient should be comfortable and his arms and head supported. His bladder should be empty. His face should be parallel with the ceiling and his neck comfortable unless there is a prominent brow, in which case the neck should be slightly extended (Figures 1–10, 1–11, 1–12). Face position is extremely important. Remember, it's easier for your patients to hyperextend their necks during surgery rather than flex them. The nonoperative eye is covered with a Fox shield so that it won't be inadvertently touched by the drapes during surgery.

I use a Chan wrist rest with a breathing bar overhead (Figures 1–13, 1–14). This keeps the drapes off the patient's face and also permits efficient oxygen delivery. The wrist rest is more appropriately called a water trough. My wrists never touch it, but it keeps the balanced saline solution (BSS) from running on to the important footpedals. The breathing bar is eliminated when the operation is rotated to the side when trying to avoid a previous filter site.

A paper drape is used before the transparent 3M incise drape. The edges of the drape should not bunch tissue centrally toward the eye. It is important to avoid creating redundancy of skin and drape at the upper lid and brow. This will interfere with phacoemulsification tip movement. The nasal edge of the paper drape should be applied to the midline bridge of the nose and no closer to the operative eye (Figure 1–15). The inferior edge is brought down and applied well away from the inferior orbital rim. The superior edge should be above the brow. The phacoemulsification handpiece will actually pivot on the brow. Extra draping material here only adds to extra height, awkwardness, and instability. The 3M incise drape is then applied so that the lashes are retracted and the eye open. Wetting the cornea just before this step helps prevent corneal abrasions. A horizontal incision is made in the drape and a vertical one above and below (Figure 1–16). An open–wire speculum is placed.

A rectus suture is placed above and below using a 4–0 silk and a non–cutting needle. A Kirby forceps is used to just grasp at the rectus muscle insertion so as not to traumatize the levator complex (Figure 1–17). The sutures are stabilized by a small hemostat above and a large Serrafine clip below. A lap towel is used to soak up the BSS.

The machine has been tested and it, along with everything else in the room, is ready (Figure 1–18).

2

Wound Construction

James A. Davison, MD

The importance of a perfect wound cannot be overstated. It represents the gateway to the entire operation. A well constructed wound can be closed by virtually any method with good results. A poorly constructed wound can be the source of astigmatism, filtration, irritation, hemorrhage, corneal trauma or worse no matter how well it's closed. A posterior scleral pocket–type incision has less of a tendency to create irritation and induce astigmatism than more anterior incisions (Figure 2–1). It also seals well and is very easily closed by virtually any suture pattern. If the wound is perfect, the suture pattern employed really does not matter much at all. The wound is so far away from the visual axis that it is difficult to induce too much with-the-rule astigmatism.

A fornix–based conjunctival mini-flap is created. To avoid unnecessary conjunctival bleeding, the scissors are driven only to the desired edge of the flap (Figure 2–2). A relaxing incision is helpful on the initiating side but is unnecessary on the other. Very light wet field cautery is applied to blanch most of the surface and deeper vessels. Tracing the larger vessels with the cautery is more efficient and less tissue destructive than sweeping from side to side (Figure 2–3).

For standard 6 mm diameter optics, a 5.5 mm scleral flap incision is created in several stages (Figure 2–4). A 6.5 mm incision is made for 7 mm optics and a 4 mm incision created for folded silicone IOL's. The eye is grasped at the 3 o'clock and 9 o'clock scleral limbus by a Penn Anderson forceps and a caliper used to mark the distance of the incision from the limbus, usually 2.5 mm to 3.0 mm (Figure 2–5). It is important to measure the distance each time because there is a tendency for "corneal creep" to occur with scleral incisions. Going a little more posterior is alright but it takes a little extra time and may create a little more bleeding. Because of the longer tunnel created, a further posterior incision may also bind the phacoemulsification tip so that it is more difficult to swivel it toward the right or left portions of the pupil.

A 57 Beaver blade is positioned so that it is at right angles to the sclera (Figure 2–6). The incision is started by drawing a line with the blade 2.5–3.0 mm from the limbus (Figures 2–7, 2–8). This blade then deepens the incision

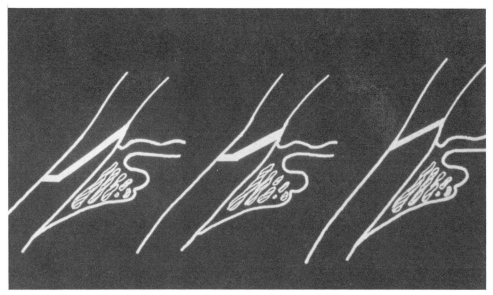

Figure 2–1. The traditional scleral pocket incision usually started 2.5 mm–3.0 mm posterior to the limbus, as in the example on the left. I now groove even deeper, about $\frac{2}{3}$ depth, then dissect forward becoming a little shallower, ultimately about $\frac{1}{2}$ depth, then form a shelf above the iris and enter the anterior chamber. Anterior or non–beveled incisions (as illustrated in the center and right) have a thinner area of contact and do not approximate as well.

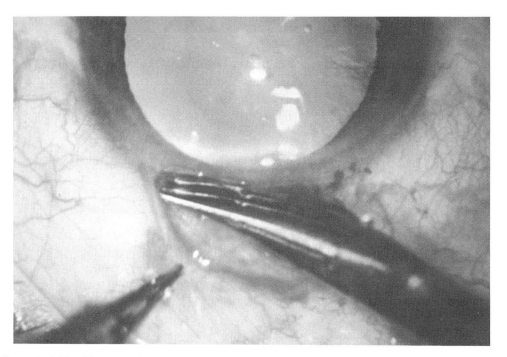

Figure 2–2. The Westcott scissors is used to create a fornix–based flap. Note that the scissors are not plunged beyond the lateral extent of the flap.

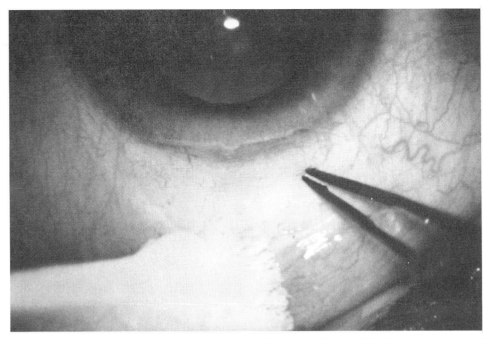

Figure 2-3. Very light bipolar cautery is applied to the sclera, "tracing" the larger, most potentially troublesome vessels.

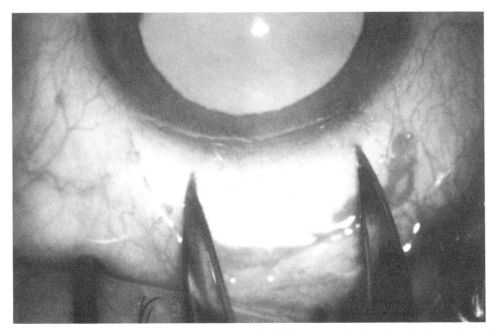

Figure 2-4. A 5.5 mm marker is used for a 6-mm IOL. Notice that while the sclera is pale, pale vessels have survived the minimally applied bipolar cautery.

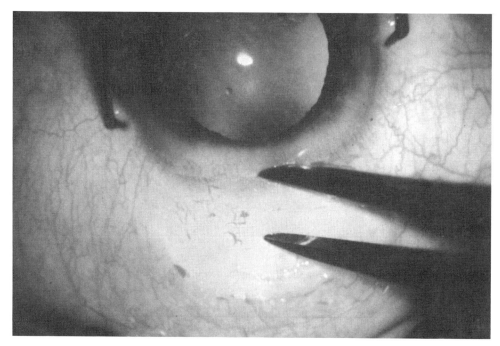

Figure 2–5. A caliper is set at 3.0 mm and used to mark distance of the incision from the conjunctival limbus.

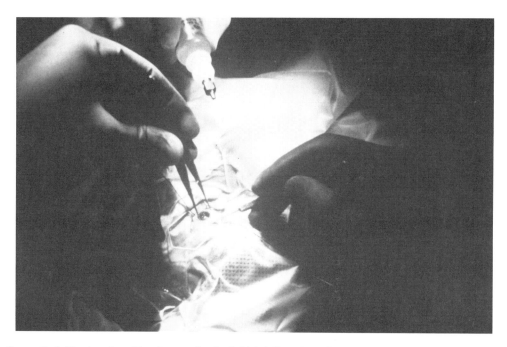

Figure 2–6. The hand position is seen for the initial delineation of the groove incision. Note that the 57 Beaver blade is at right angles to the sclera. The eye is fixated with Penn Anderson forceps. A caliper has been used to make an impression 3.0 mm from the limbus. This is the distance of the incision from the limbus.

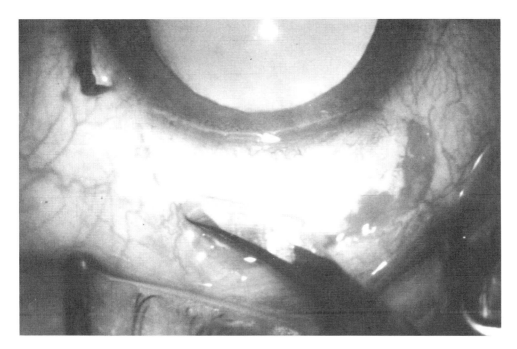

Figure 2–7. The initial groove is created with the 57 Beaver blade. Very little pressure is applied. Concentrate only on drawing a line. Do not think about creating a depth to the incision at this point.

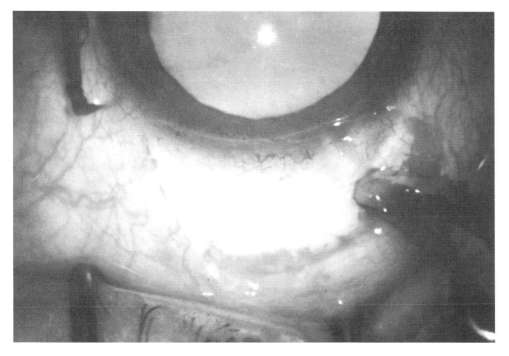

Figure 2–8. The initial curved scratch has been completed.

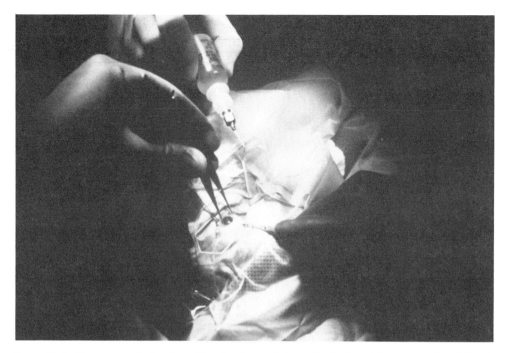

Figure 2–9. The surgeon's hand position is changed dramatically to achieve an undermining in a scleral plane parallel incision. If this is not done, premature entry will occur. Good irrigation is important.

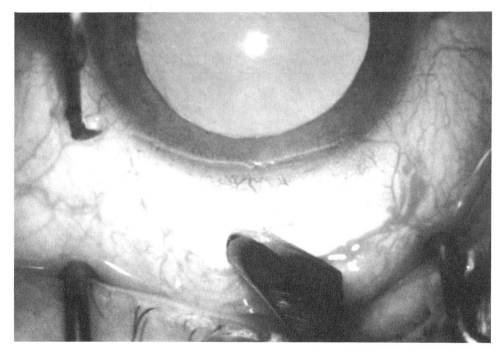

Figure 2–10. The blade is brought back in this new flat angle.

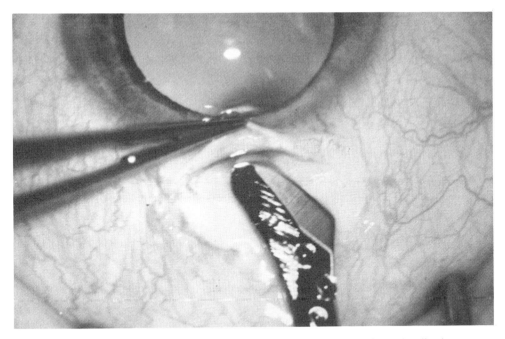

Figure 2–11. Scleral dissection proceeds with the 57 Beaver blade under direct visualization.

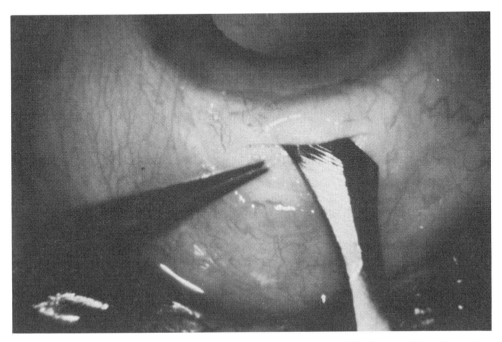

Figure 2–12. The blade's contour can be seen through the sclera early incision making. Direct visualization of the developing tissue plane is not necessary at all times, but when in doubt, look.

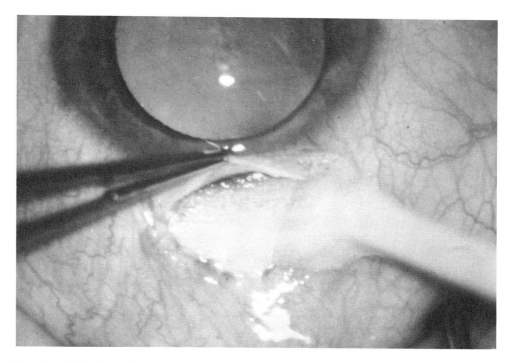

Figure 2–13. The depth of the scleral dissection can be monitored more accurately by taking frequent looks at the scleral fibers and wound after they have been dried.

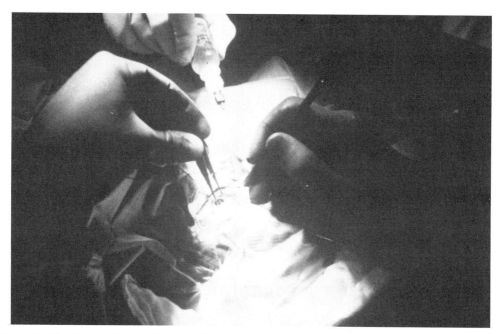

Figure 2–14. A Grieshaber angled round blade is used to create the rest of the scleral tunnel. Note that this blade can be held like a pencil. This is important if the eye cannot be rotated down or if the brow is especially prominent.

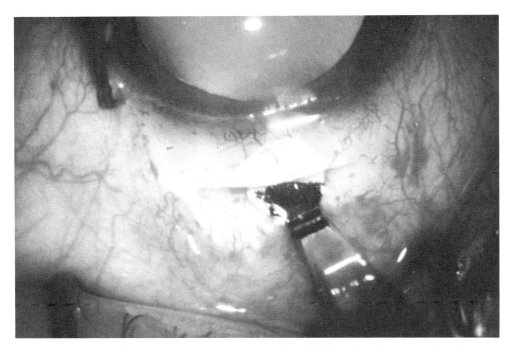

Figure 2–15. The pocket has almost been completed. Note that the blade is easily visible through the partial–thickness scleral flap.

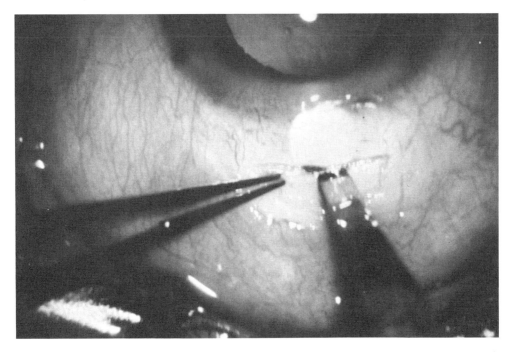

Figure 2–16. The handle of the round blade is twisted in the surgeon's fingers to produce a slow oscillating motion. The forward drive of this blade is stopped when its leading edge just passes the vascular arcade.

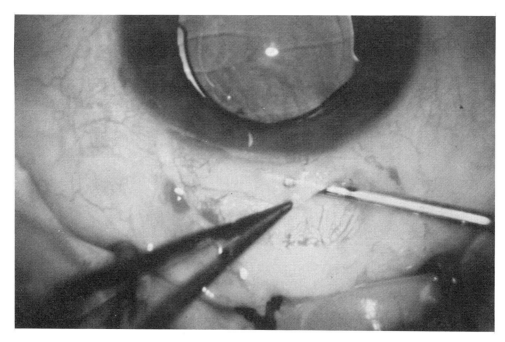

Figure 2–17. A very thin flap has been created and an actual buttonhole produced by being too shallow. Careful suturing will be required to bridge over the thin flap and anchor to tissue of substantial substance.

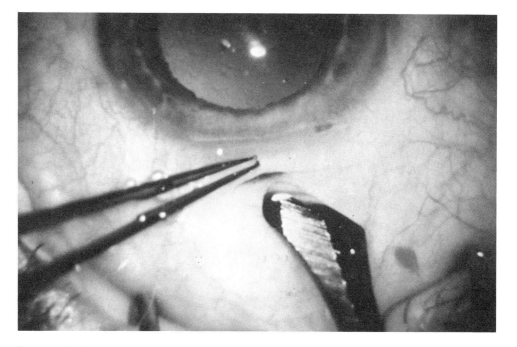

Figure 2–18. The dissection of the scleral flap has been too steep relative to the curvature of the sclera and the cornea. Premature entry has occurred from approaching the limbus "too hot." The incision should have been placed 0.5 mm further back and the appropriate tissue plane identified for dissection.

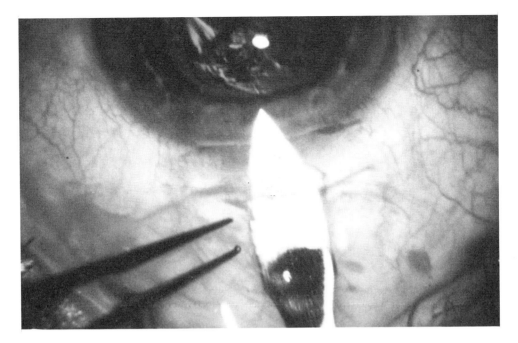

Figure 2–19. After the anterior capsulotomy, the keratome widens the incision centrally at its thinnest part. The full–thickness width of the incision should be measured. If it is greater than 3 mm, a 10–0 nylon suture should be placed to make the wound smaller.

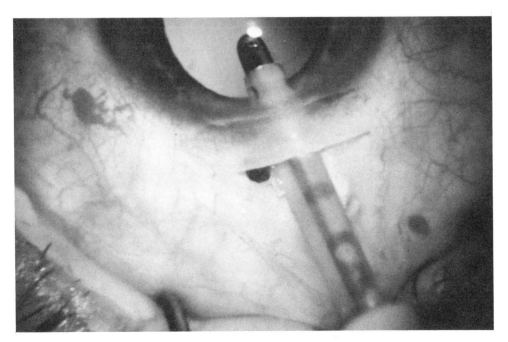

Figure 2–20. A suture was not placed, but should have been. Relatively minor iris prolapse is observed. Generally, this should be repaired immediately or substantial depigmentation, sphincter rupture, or bleeding may occur.

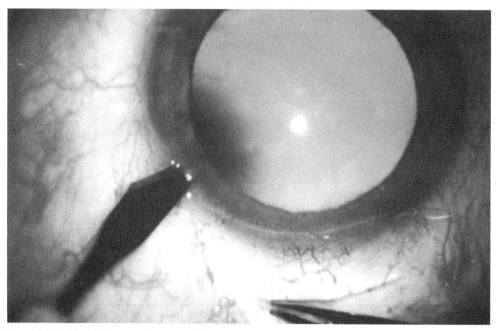

Figure 2–21. The .12 forceps is grasped in the right hand and full thickness entry accomplished into the anterior chamber with a 22.5–degree disposable super blade. The entry is not parallel with the iris but aims at the pars plana 180 degrees away.

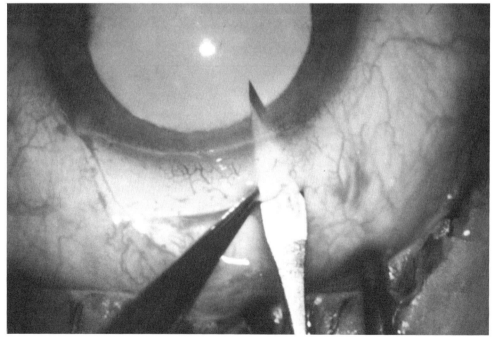

Figure 2–22. An iris parallel entry is made through the remaining fibers at the right–hand apex of the scleral pocket incision.

slightly as the beveling process is begun by changing the angle of the blade so that it is more parallel with the sclera and drawing the blade back to the origin of the initial scratch (Figures 2–9, 2–10).

In most circumstances, the incision will be best completed by dissecting the sclera forward with the 57 blade under direct visualization (Figures 2–11, 2–12, 2–13). A dry sclera makes it easier to see the individual fibers during the dissection. If the brow is steep or the eye cannot be rotated downward easily, an angled Grieshaber round blade can be used to dissect the sclera forward (Figures 2–14, 2–15). The Grieshaber blade often can be used to tailor the last portion of the incision as it emerges into clear cornea and to prevent premature entry.

The anterior dissection should not be stopped until reaching just beyond the limbal vessels (Figure 2–16). This way, premature entry—with its troublesome iris prolapse or Descemet's disruptions will not occur.

Problems can occur when creating the wound. The two most common problems are not grooving deep enough and angling in "too hot" toward Descemet's. The former leads to initiating too thin a flap (Figure 2–17) and the latter creates premature entry (Figures 2–18, 2–19, 2–20). Defective thin flaps and troublesome iris prolapse and their sequelae are the results of these common errors. The tendency to be too shallow or too deep can be prevented in part by remembering that the radius of curvature of the cornea is shorter than that of the sclera with the transition occurring at the limbus.

I consciously try to become a little shallow with the round blade. The blade's convex posterior surface helps me do this. I have not run into the ciliary body yet, but great care must be taken not to go too deep in highly myopic eyes with thin sclera or to strip Descemet's with too central an interior entry in the smaller hyperopic eye. Extremes in globe dimension and scleral thickness are two of the chief sources of special challenge to the ophthalmic surgeon, not only during wound construction, but throughout the operation.

The eye is entered with a 22.5–degree disposable blade just on the conjunctival limbus at the 1:30 o'clock position (Figure 2–21). A polished down 0.5 mm cyclodialysis spatula (to about 0.4 mm) and 30–gauge blunt cannulas will be introduced here. The eye is also entered with this blade approximately 1.5 mm from the right hand wound extreme. Care must be taken to follow already created wound plane and also be parallel with the iris at this entry (Figure 2–22). A 22–gauge blunt cannula will be introduced through this stab incision to instill a viscoelastic substance. A 3.0 mm keratome (Beaver, 5520) will later enlarge this incision for the ultrasonic tip. Both incisions should enter the anterior chamber well above the iris plane to prevent iris prolapse. The second incision should be larger than the first so it can accommodate the 22–gauge cannula.

Section II
THE CAPSULORHEXIS

Introduction to Capsulorhexis

Paul S. Koch, MD

Making a continuous tear capsulotomy through capsulorhexis, ripping of the anterior capsule is, in my opinion, the single most important development leading to the advanced phacoemulsification techniques. It completely changes the way we think about phacoemulsification and the way we choose to perform it.

It's funny, but when it was introduced it seems to have been lost in an odd set of marketing blunders. It should have achieved general acceptance years before many people picked up on it.

I was using some of the STAAR sulcus lenses, and, like a lot of people, after I put in a few dozen of them I decided I wasn't really all that happy with them. The company came back to me with their bag lens, which had the unfortunate reputation of slipping out of the bag, so I wasn't interested.

They came back a few months later and showed me a videotape about capsulorhexis, as described by Drs. Gimbel and Neuhann. They told me that if I would switch to their technique, I would be able to use some more of their lenses. Stupidly, I looked at the capsulorhexis as a marketing ploy and did not recognize its true value. I declined to even try it.

Looking back now from my new vantage point, I think that if they had argued that capsulorhexis was the single best way to perform an anterior capsulotomy, and if I were clever enough to realize they were right, I would have begun doing it years ago. What I've learned since then is that capsulorhexis has changed the face of cataract surgery.

Capsulorhexis, continuous capsulotomy, forms an incredibly strong opening in the anterior capsule. In most patients you can bend and twist the anterior capsule, but you cannot rip it.

The capsular bag takes on a different perspective when you perform capsulorhexis. For all practical purposes, the capsular bag remains intact, just like it was before the capsulotomy. You haven't ripped out a ragged opening in the anterior capsule, an opening that can enlarge, stretch, or rip apart. You've simply punched out an opening while preserving most of the normal shape and strength of the bag.

With a can-opener capsulotomy you can push the nucleus halfway across the eye without peeling off zonules because the capsulotomy stretches open to accomodate it. In fact, that was the original Kelman method for dislocating the nucleus into the anterior chamber. After the capsulotomy, the nucleus was engaged with the cystotome. It was moved halfway across the eye until the equator passed the midline of the pupil. The nucleus then could be prolapsed into the anterior chamber.

You can't do that with a capsulorhexis. The anterior half of the capsule will not stretch and accommodate the pressures exerted by the nucleus because, for all practical purposes, the bag is intact. Move the nucleus two or three millimeters and you will start separating zonules. It's like moving the intact cataract, not just the nucleus.

This means that capsulorhexis is not a great capsulotomy for nucleus expression procedures. Unless you make a very large one, you could accidentally express the entire cataract because the nucleus will remain trapped within the bag.

The capsulorhexis is designed to maintain the integrity of the capsular bag, and so it should be considered as no more than a hole punched in the surface of the cataract through which you can pass instrument tips to work on the cataract. It should not be considered as an opening through which the intact nucleus can be removed.

Clearly, if you have a small nucleus and a large capsulorhexis you can perform a nucleus expression procedure, but that's not what we're talking about here. This discussion is about a "normal" capsulorhexis, measuring about five millimeters, and a "normal" cataractous nucleus.

Because the capsulorhexis is designed to keep everything in the bag, you have to perform all of your phacoemulsification in the bag. Forget about nucleus prolapse and iris plane phacoemulsification. That's another technique for another capsulotomy. Once you perform capsulorhexis, you should plan to perform your procedure in the bag.

In the iris plane procedures, you want the nucleus held back in the capsular bag during sculpting, both for control and to keep your tip away from the cornea. Because the can-opener capsulotomy did not do this, it was advised to maintain the cortical adhesions until the end of this step. After sculpting was completed, you could loosen the nucleus physically or hydraulically, prolapse the nucleus, and continue.

Now it's just the opposite. The nearly intact anterior capsule holds the nucleus in the capsular bag so you can loosen it all you want. We'll spend some time on hydrodissection and hydrodelamination in the next section.

Now you may see why this is an advanced technique. You cannot convert part way through the case to a nucleus expression procedure. Your capsulotomy prevents you from prolapsing the nucleus. The only way you could make a conversion is by going back and revising the capsulotomy, making relaxing incisions at 10:30 and at 1:30. But face it, if you're bummed out enough to want to convert, you're probably too bummed out to want to go back and revise or reconstruct a capsulotomy.

Three chapters that follow demonstrate two methods of performing capsulorhexis. In the first one, Rob Kershner discusses the embryologic and anatomic factors that need to be considered in evaluating and performing capsulorhexis. He then describes the needle technique of performing one. Next is the forceps technique, which I find more comfortable. Both are good techniques, and I am unaware of any factor that would make one better than the other except for surgeon preference. Finally, Jim Davison presents a detailed analysis of the different types and patterns of capsulorhexis.

3

Embryology, Anatomy and Needle Capsulotomy

Robert M. Kershner, MD

For most ophthalmologists, the classic teaching of the capsulotomy was to perform whatever method allowed the quickest way of getting to the cataract. As phacoemulsification technology improved, the operation changed. We developed smaller incisions, endocapsular and in situ phacoemulsification techniques and small–incision intraocular lenses. The manner in which the capsulotomy was performed became more critical with these developments.

There are as many methods to perform capsulotomy as there are surgeons who perform surgery. However, among advanced phacoemulsification surgeons, smooth, continuous tear capsulotomies are gaining popularity as the procedure of choice. This chapter will discuss a simple method of performing capsulorhexis to achieve a smooth-edged small and central capsulotomy that on the one hand allows access to the cataract for phacoemulsification, irrigation, and aspiration, and on the other hand facilitates in–the–bag implantation of the intraocular lens.

Embryology and Anatomy

Before we take a look at why capsulorhexis is the procedure of choice, we should explore the anatomical and embryologic reasons why retaining a large rim of anterior capsule is desirable. The human crystalline lens is developed from surface ectoderm by induction from the lens placode. Invagination of the surface ectoderm creates the primitive lens. As a result, the surface ectodermal cells are arranged inside-out with the basal portion of the cell oriented outward. As the basement membrane of these cells creates a second basal lamina, the lens capsule is formed (Figure 3–1).

Like its analog, the skin, the lens grows throughout life and the capsule along with it. The lens epithelial cells reside on the anterior surface and extend to the equatorial zone of the lens capsule. The cells that are located centrally probably do not undergo much mitotic activity. However, the peripheral lens epithelial cells actively divide and give rise to the developing lens fibers. This activity creates a lens capsule that is roughly twice as thick anteriorly as it is posteriorly (Figure 3–2).

Although the lens capsule appears under the microscope to be a single layer, it is actually made up of approximately forty lamellae. One of the important functions of this structure is to provide support for the zonules, which hold the lens in place (Figure 3–3).

Chemically, the capsule is made up of insoluble protein that is similar to collagen, with a small amount of polysaccharide material. Although the anterior and posterior diameter of the lens varies, the capsule plays little, if any, role in molding the shape of the lens. The overall diameter of the lens has remarkable little variation, being approximately 10 mm (Figure 3–4).

The Continuous Tear Capsulotomy

It has been taught that, for the beginning phaco surgeon, the ideal capsulotomy is the largest. This is because the maneuvers of central sculpting, nuclear rotation, and prolapsing of the residual cortical rim from within the capsular bag requires a large area of access to the cataract. Smaller capsulotomies can be an impediment to the two-handed phaco surgeon who wishes to elevate the nucleus to the iris plane for its final removal.

Can-opener capsulotomy techniques create multiple cuts in the anterior lens capsule, which can lead to inadvertent tears and tags that can interfere with phacoemulsification and aspiration (Figure 3–5). The notches that can-opener–type capsulotomies create can be the source of inadvertent extension of tears to the posterior capsule.

A smooth, continuous tear capsulotomy affords the surgeon the advantage of a single continuous edge. This reduces the risk of capsular tears extending through the equator and provides a resilience to the capsule during phacoemulsification and lens implantation (Figure 3–6).

What is the best shape for the anterior capsulotomy? Round capsulotomies would logically appear to be best for round intraocular lens optics. They contract circumferentially and equally, facilitating centration of the IOL. Ovoid or elliptical capsulotomies can be positioned to allow access to the superior cortex at 12:00 o'clock, making the removal of this most difficult area simpler to perform during aspiration. Ovoid capsulotomies, however, may limit the size and shape of the implant optics that can be used. Unless the edges of the optic are covered, especially with smaller optic implants, edge glare can occur. Ovoid or decentered capsulotomies that do not cover optic edges may necessitate larger optic implants, defeating some of the advantages of the newer procedures. Logic dictates, then, that round is probably the best shape.

What, then, is the best size? Most popular IOL optics are 6 mm in diameter.

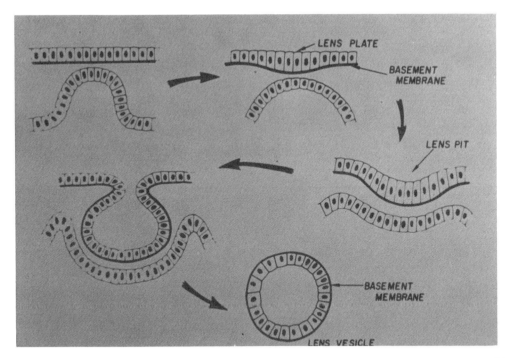

Figure 3–1. The human lens is developed from surface ectoderm. These are arranged inside-out with the basal portion of the cell oriented outward. As the basement membrane of these cells create a second basal lamina, the lens capsule is formed.

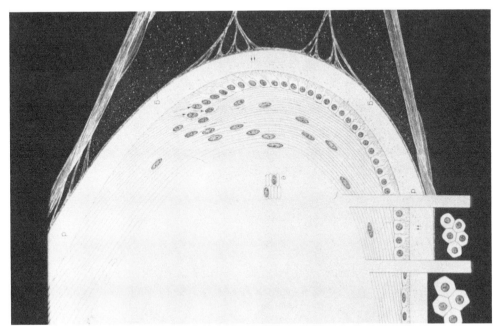

Figure 3–2. The lens epithelial cells reside on the anterior lens capsule. Mitotic activity of these cells creates a lens capsule that is roughly twice as thick anteriorly as it is posteriorly.

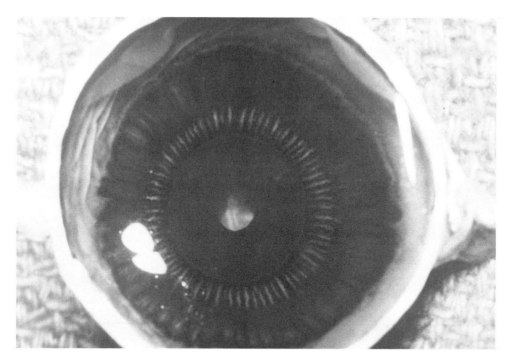

Figure 3-3. While the lens capsule appears to be a single layer, it is made up of approximately 40 lamellae. One function of the capsule is to support the zonules.

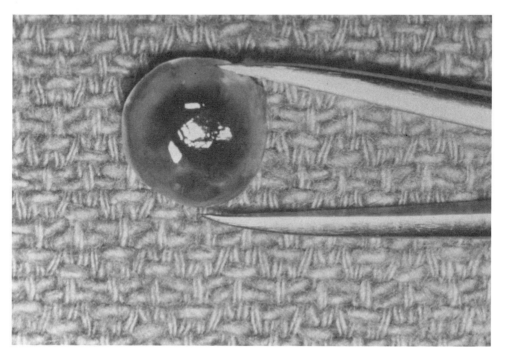

Figure 3-4. The normal lens measures about 10 mm.

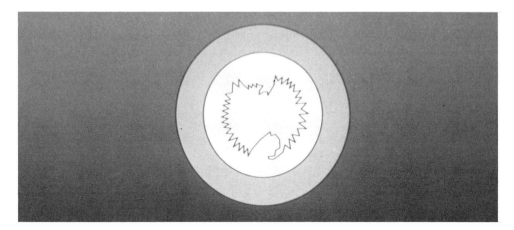

Figure 3–5. Large capsulotomy avoids tags that interfere with emulsification and aspiration.

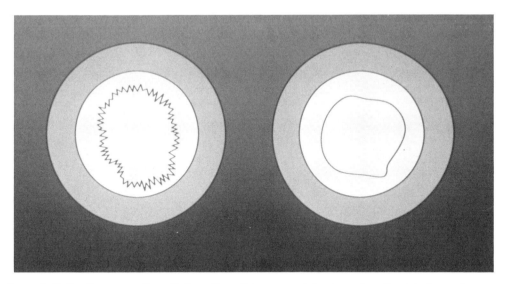

Figure 3–6. Can Opener vs. Smooth Tear. Smooth tear capsulotomies reduce the risk of capsular tears extending through the equator and provides resilience to the capsule during phacoemulsification and lens implantation.

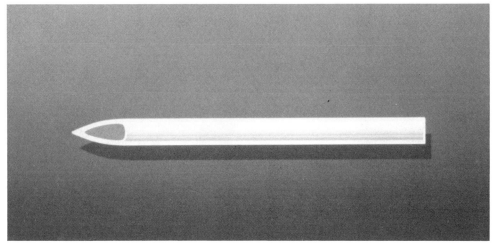

Figure 3–7. A needle capsulotomy begins with a 27–gauge ⅜″ needle.

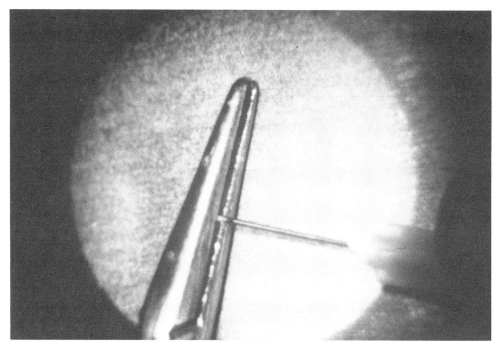

Figure 3–8. The needle is grasped in a needle holder and bent two ways.

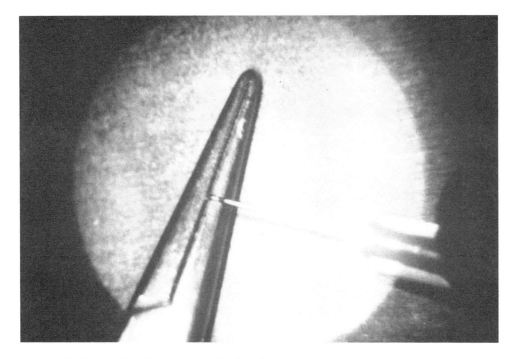

Figures 3–9. The first bend is away from the bevel.

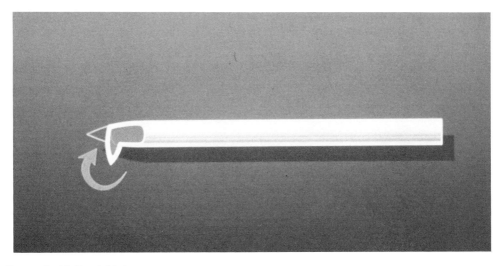

Figures 3–10. Bend "tip" of needle 90°.

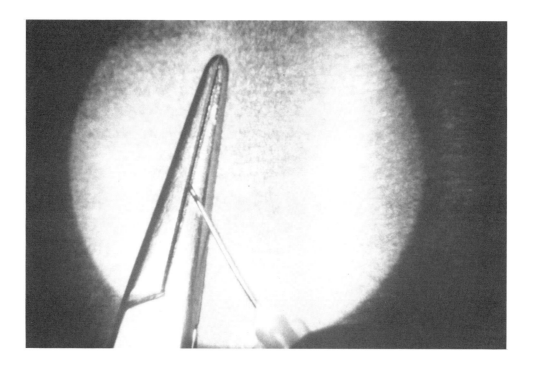

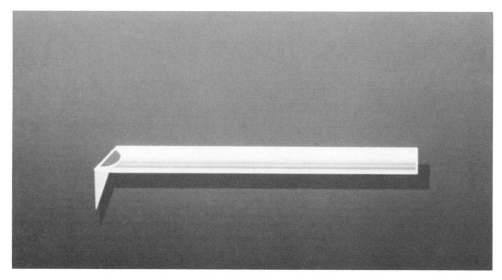

Figures 3–11, 3–12. The second bend, while still in the needle holder, is 45 degrees to the side.

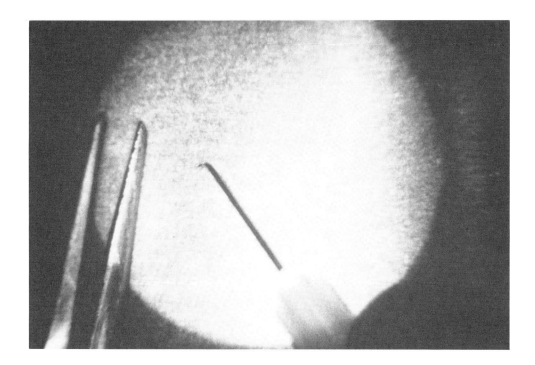

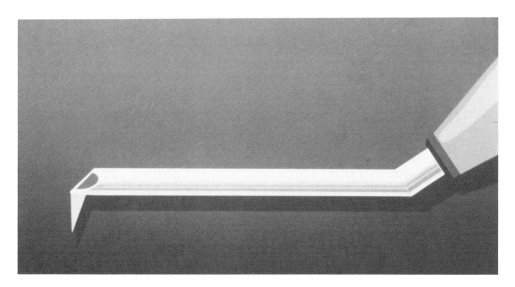

Figures 3–13, 3–14. After the tip is bent, the shaft is bent 30 degrees at the hub and connected to the viscoelastic syringe.

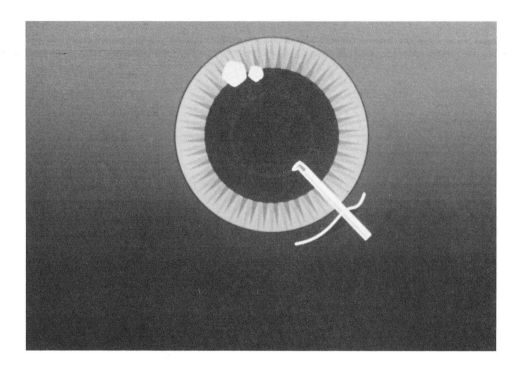

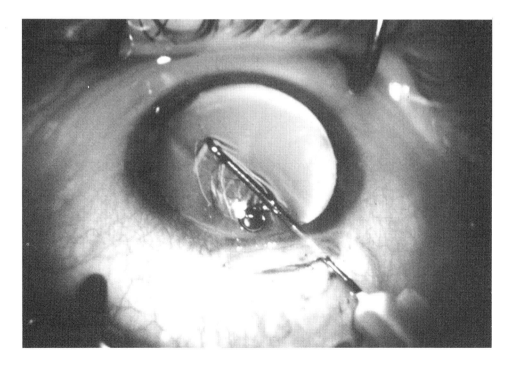

Figures 3–15, 3–16. The first motion is a snag of the capsule at 12:00 o'clock. This starts the tear in a clockwise fashion.

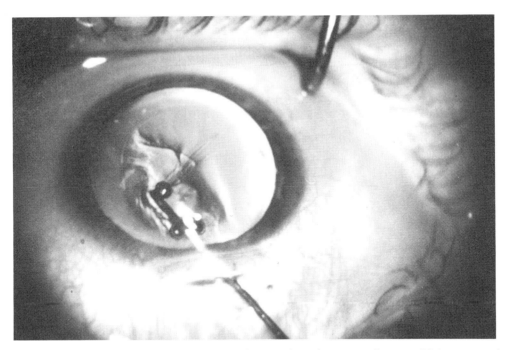

Figure 3–17. The needle continues to re-engage the capsule as the tear continues until it is connected to the original incision.

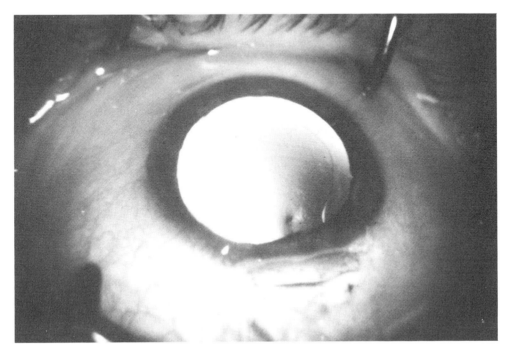

Figure 3–18. The result is a round, smooth-edged, continuous tear capsulorhexis.

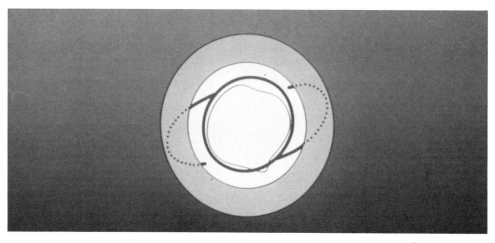

Figure 3–19. A small capsulotomy simplifies and guarantees in-the-bag lens placement.

New ones are 5 mm. Capsulotomies larger than 5 mm would allow exposure of an IOL edge and be less likely to sequester the lens within the capsular bag. Capsulotomies smaller than 4 mm make phacoemulsification more difficult and may prevent implantation of intraocular lenses with optics larger than 6 mm. The ideal diameter of the anterior capsulotomy, then, is probably somewhere between 4 mm and 5 mm.

Making the Continuous Tear Capsulotomy

There are several ways in which to create a smooth, continuous tear capsulotomy. Several automated devices have been advanced recently, and many of them appear to work well. Automated capsulotomy devices, however, carry with them the additional expense and need to have additional instrumentation. The two most commonly used methods for capsulorhexis are the bent needle and the capsular forceps technique. The Utrata forceps allows a controlled capsular tear. There is, however, a learning curve associated with the use of this instrument, and it does require the surgeon to have one additional instrument on the field. For many surgeons, a bent needle is the simplest way to learn and master smooth, continuous tear capsulotomy. (see Figures 3–7 through 3–15).

A 27-gauge needle can be bent with a needleholder close to the end of the needle tip 90 degrees, and then bent again 45 degrees, this time in the plane of the needle shaft. This creates both a cutting and pushing edge on the needle at its tip, permitting one to direct where to tear the capsule. The bent 27-guage needle then can be connected directly to the viscoelastic syringe. Once the incision into the eye has been made, the capsulotomy needle can be introduced and a bolus of viscoelastic placed in the anterior chamber.

The first motion is a snag of the capsule at 12:00 o'clock. This starts the tear in a clockwise fashion. Once the tear is created at 1:00 o'clock, the tip of the needle is then used to engage the flap of anterior capsule and gently fold it over the underlying capsule as the needle is pushed in a clockwise direction. As the capsulotomy extends, directing the needle toward the center of the lens directs the tear in a circular fashion. The needle is then used to re-engage the capsule as the tear continues until it is connected to the original incision. The result is a round, smooth-edged, continuous tear capsulorhexis. Phacoemulsification can then be performed entirely within the capsular bag through this small central capsulotomy (Figures 3-16, 3-17, and 3-18).

Retaining a large section of the anterior capsule provides additional support should a posterior tear inadvertently be created during phacoemulsification or irrigation and aspiration. The anterior capsular support can then be utilized for sulcus placement of the intraocular lens on top of the anterior capsular leaflet.

The capsulotomy will vary by the size of the pupil. Smaller pupils require a down-scaling of the capsulorhexis. In addition, greater than 7 mm intraocular lens optics may require slightly larger capsulotomies.

If we take the anatomical features of the lens capsule into account, we can direct our approach to the capsulotomy to derive the most benefit from these features. First, because the anterior lens capsule is thicker, it provides more structural support to the capsular bag. This allows more support to the lens optic and haptic placement, preventing inadvertent movement of the lens and IOL decentrations.

Second, the lens epithelial cells that remain following irrigation and aspiration of the capsular bag will ultimately undergo metaplasia. These cells cause fibrosis and scarring of the lens capsule and subsequent contracture. While frequently thought of as an undesirable activity, this is helpful in securing the intraocular lens within the capsular bag.

With large capsulotomies, the remaining lens epithelial cells have no place to go but posteriorly, where their metaplasia interferes with visual function. However, the smaller anterior capsulotomy allows the lens epithelial cells to remain on the anterior lens capsule. There, they facilitate fibrosis of the capsule, creating a circumferential or sphincter-like effect surrounding the capsulotomy. Microscopic examination shows these epithelial cells to migrate to the cut edge of the capsulotomy and no further. It is there that contracture will occur. It is this contracture of the small central capsulotomy that sequesters the intraocular lens within the bag (Figure 3-19).

I reviewed more than 1,000 cases comparing an equal number of can-opener versus capsulorhexis capsulotomies (see Tables 3-1 and 3-2). The incidence of lens decentration, capsular opacification, and inadvertent rupture of the capsule was far less when capsulorhexis was used for the anterior capsulotomy.

Small, smooth-edged continuous tear capsulotomies become necessary with the use of high technology IOL's, which require more critical centration, and

TABLE 3–1.   In 1,000 cases of can-opener capsulotomy, the incidence of lens decentration was 16%, capsular opacification 29%, and capsular rupture 8%.

1000 Cases Can-Opener Capsulotomy

Lens Decentration	16%
Capsular Opacification	29%
Capsular Rupture	8%

TABLE 3–2.   In 1,000 cases of capsulorhexis, the incidence of lens decentration was 3%, capsular opacification 6%, and capsular rupture 4%.

1000 Cases Capsulorhexis Capsulotomy

Lens Decentration	3%
Capsular Opacification	6%
Capsular Rupture	4%

and smaller soft, foldable, and injectable IOL's, which require sequestering within the capsular bag.

Studies by others (Patel and Apple) have shown that maintenance of the anterior capsule provides additional protection to the endothelium during phacoemulsification. This, in combination with sequestering the intraocular lens within the capsular bag to reduce the biological contact with reactive tissues, potentially reduces intraocular inflammation, reducing the occurrence of capsular opacification.

These distinct advantages make capsulorhexis superior to can-opener style capsulotomies and further strengthen the need to master this technique as part of the evolution to advanced phacoemulsification technology.

4

Forceps Capsulotomy
Paul S. Koch, MD

Forceps Capsulorhexis

I prefer to perform a continuous tear capsulotomy with forceps because I find it easier. I do not necessarily think it is any better than using a needle—the results are the same. I just think it is easier to grasp the capsule and pull it around than it is to nudge it with a needle.

Capsulotomy forceps are long tipped forceps, much like Kelman-McPherson forceps, but with a pair of teeth pointing down at the tip. The main examples are the Utrata forceps and variants like the Kraff-Utrata forceps and the Kershner forceps, which has very sharp teeth so it can be used as either a sharp instrument for needle capsulorhexis or as forceps for this technique.

The preparation for a forceps capsulotomy is the making of a 3 mm incision, the same one that will be used for the phacoemulsification. This is followed by the injection of a viscoelastic. I do not think a forceps capsulotomy can be performed without it.

Before grasping the capsule with a forceps, however, it is necessary to form a flap. I do this with an old-fashioned blunt Kelman cystotome. I place it in the middle of the anterior capsule, just engaging its surface, and then tug it towards the incision, forming a mini Christmas Tree capsulotomy.

The small flap of anterior capsule is wiped down onto the more superior anterior capsule, flattening the flap on intact capsule. It is important to have the capsular flap down flat at all times in order to control the tear. When the flap is initially laid flat, it's like a triangle pointing toward the incision.

If the cystotome cuts into the capsule and makes a linear incision, any attempt to grasp the edge of the capsule and tear it will result in continuation of the linear tear to the equator of the lens. The better thing to do is push sideways on the incision with the broad side profile of the cystotome as the incision is being made. It will direct the cut to the side, beginning a curved tear in the direction of the push. After the curved tear is begun, the forceps can be used to continue it.

Figure 4–1. A blunt Kelman cystotome is placed in the eye just below the center of the pupil, and tne anterior capsule is lightly engaged.

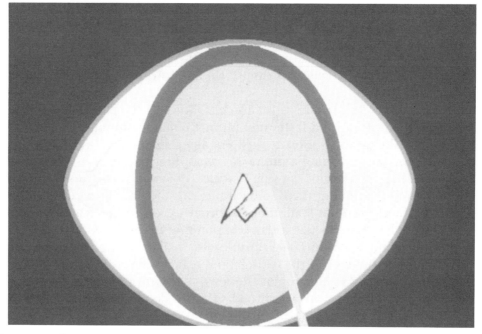

Figure 4–2. The anterior capsule is engaged and pulled toward the incision, creating a triangular Christmas tree flap.

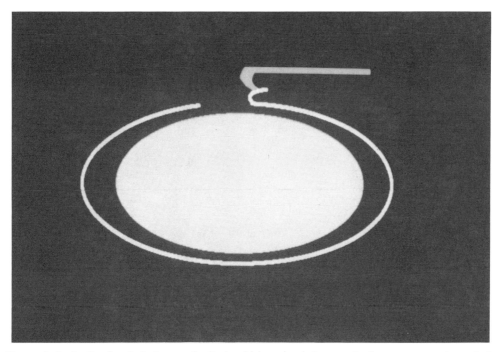

Figure 4–3. As the flap is being made, it should be wiped down onto the more superior anterior capsule, flattening it.

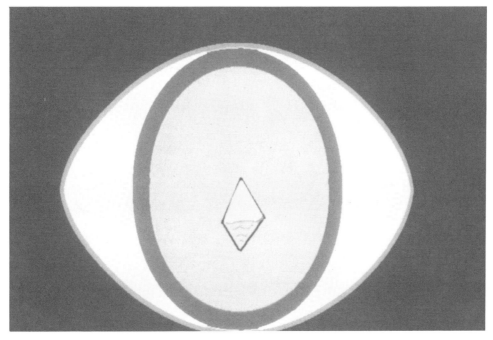

Figure 4–4. When the capsular flap is laid flat, it is like a triangle pointing toward the cornea.

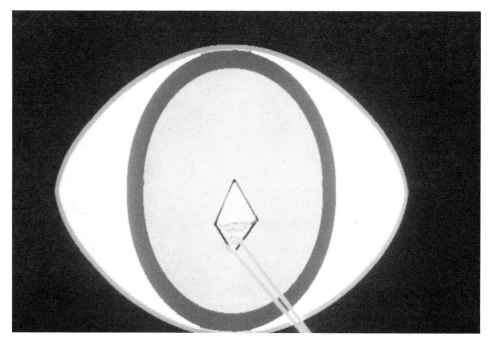

Figure 4–5. The capsulorhexis forceps are used to grasp the capsular flap and tear it around in a counter-clockwise fashion.

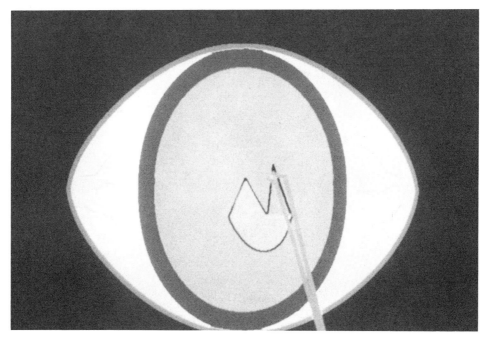

Figure 4–6. When the tear has gone three clock hours, it is time to stop and regrasp the flap again.

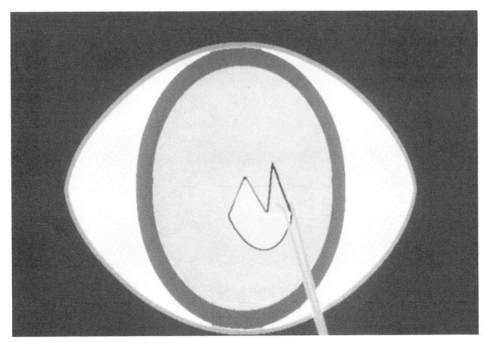

Figure 4–7. The capsular flap is regrasped closed to its base before another three clock hours are torn.

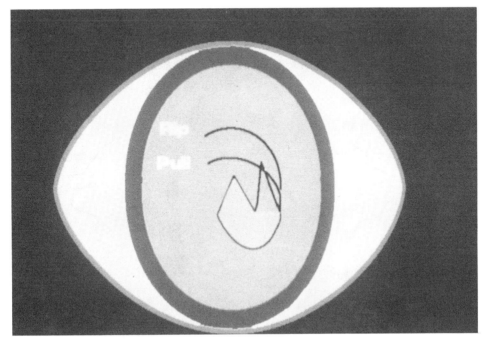

Figure 4–8. The tear begins with a slight forward motion to get it started. The movement of the forceps is then in a radius within the path that the tear will eventually follow.

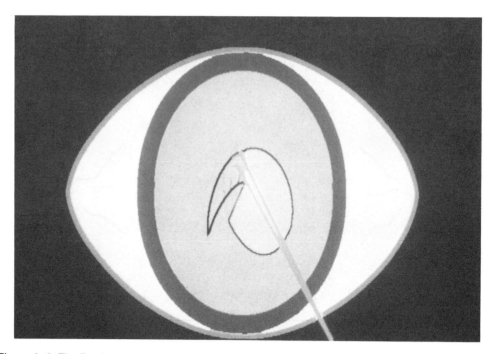

Figure 4–9. The flap is regrasped at the six o'clock position and torn three more clock hours.

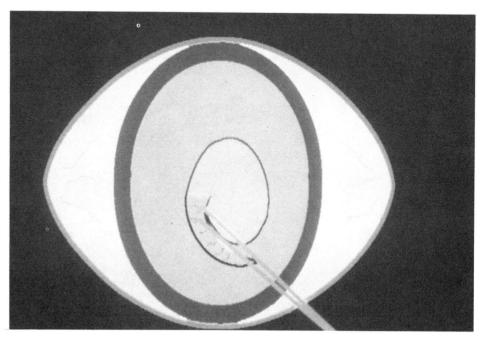

Figure 4–10. The flap is regrasped at the three o'clock position in preparation for the final tear.

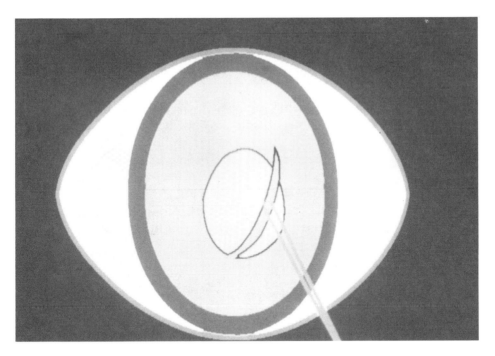

Figure 4–11. The final pull of the flap is toward the seven o'clock position, completing the continuous tear capsulotomy

The forceps are first used to grasp the capsule at one edge of the tear, the left hand edge if the tear is designed to go counterclockwise. The flap is pulled a little bit toward the incision and then swept up towards the 9:00 o'clock position. The smooth tear in the capsule can be seen to follow the tip, making a tear with a slightly larger radius than the motion of the forceps.

The capsule needs to be regrasped every three clock hours, more often if necessary. Attempts to pull the capsule more than three hours with a single grasp often leads to the capsule going off course and heading for the equator.

Once the capsule is regrasped, it is torn again toward 6:00 o'clock. The radius of the forceps is smaller than the radius desired for the capsular tear. After three more clock hours of tear, the flap is flattened on the intact anterior capsule and then regrasped near the cut edge of capsule.

Another pass brings the capsulotomy toward 3:00 o'clock, and it is then regrasped and torn toward 12:00 o'clock. As the circle nears completion, the capsule is directed back inferiorly, toward 7:00 o'clock or so, causing the flap to complete the circle and snap off near the point where the curved portion of the tear was originally begun. With any luck, the capsulotomy will be fairly round, centered, and smooth.

If the Capsulotomy Goes off Course

If the capsulotomy goes off course, there are ways to recover it. The most common places where it goes off course are at 10:00 o'clock, during the first tear, and at 5:00 o'clock, as you make the inferior turn and head back up toward the incision.

If it goes off course at 10:00 o'clock, it is relatively easy to use small capsulotomy scissors to cut the capsule and begin a new flap, which can be grasped with the forceps and torn toward 6:00 o'clock.

If it goes off course at 5:00 o'clock, it's a tougher problem. Long scissors have been designed to reach across the anterior chamber and begin a new tear, but these are often unavailable or unwieldy. Another approach is to use scissors to begin a new flap just inside the incision, where the curved tear began, only in the other direction. The final tactic is to abandon the continuous tear capsulotomy and finish with a can-opener.

5

Capsulotomy Tear Patterns

James A. Davison, MD

Anterior Capsulotomy Tear Patterns

Anterior capsulotomies should be central and circular to provide the most favorable optical conditions and important structural symmetry both early after the operation and years later. The perfect size of the anterior capsulotomy is open to debate. I think, though, that it should either cover or not cover the entire peripheral optic in symmetrical fashion for 360 degrees. My preference is to slightly overlap the anterior IOL optic with approximately 0.25 mm of the anterior capsular remnant.

If the capsulotomy is too small, surgery will be unnecessarily difficult, resulting in increased risk to the anterior capsular remnant and posterior capsule as well. Too small a capsulotomy can also cause later optical problems. As capsular fibrosis occurs, the size of the opening in the anterior capsule may stay the same as originally created or it may get a little smaller (Figures 5–1, 5–2). It won't ever get larger.

The problem of anterior capsulotomy shrinkage may actually be made worse by capsulorhexis because of a sphincter-like effect that appears to develop as the anterior capsular remnant undergoes the contraction of epithelial cell fibrosis (Figure 5–3). It can become incredibly small in rare cases of pseudoexfoliation even if it was created with the correct dimension. The forces of progressive capsular fibrosis can drastically shrink the opening because of the lack of counteracting zonular traction in these cases (Figure 5–4).

Ultrasonic removal or cryopexy of the anterior capsular epithelial cells in pseudoexfoliation cases might be particularly helpful in preventing this complication, which will ultimately end in a completely dislocated IOL within its detached capsular bag. Cryopexy of these cells also may reduce the number of cases in which an exaggerated epithelial cell response creates a proteinaceous glue–like sheen at the edge of the capsulorhexis, which can further reduce the functional visual area of the IOL optic (Figure 5–5).

If approximately one-quarter or more of the optic is left uncovered, the edge of the capsulorhexis may hammock the optic edge and slightly decenter it as capsular fibrosis occurs. This subtle late decentering tendency may be exaggerated if the IOL haptics are soft but, in most cases, it will still not be clinically significant. It may be easier to compound the conditions necessary for this type of decentration and increase its clinical significance by introducing the asymmetric, or non-circular optic.

As a general rule, I believe that the anterior capsule should overlap the peripheral optic about 0.25 mm all the way around its perimeter at the time of surgery. If a 6.0 mm optic IOL is being used, a 5.5 mm continuous tear circular capsulotomy should be fashioned. A 6.5 mm capsulotomy is created when a 7.0 mm IOL optic is used. Capsular fibrosis will generally increase this coverage by another 0.25 to 0.5 mm. If the capsulorhexis is the correct size at surgery, this reduction in size secondary to fibrotic contraction is usually not optically significant. Perfect capsular openings are more critical in younger patients or those in whom multifocal lenses have been implanted. Some of the advantages of a 7 mm lens will be lost if the anterior capsular opening is only 4.5 mm to 5.0 mm in diameter. The peripheral focusing zone of a multifocal lens can be rendered totally noncontributory by too small a capsular opening (Figure 5–6).

Creating the Anterior Capsulotomy

The rectus sutures should be released while creating the anterior capsulotomy. The anterior chamber should then be filled with enough viscoelastic to maintain a taut capsular surface (Figures 5–7, 5–8, and 5–9). This pressure will prevent troublesome wrinkles that might develop while the capsulotomy is being scribed. The capsulotomy is begun with a puncture of the anterior capsule by the 25–gauge needle, which has been bent a sharp 90 degrees at the tip and at about 30 degrees at its insertion into the hub (Figures 5–10, 5–11). Four basic tear patterns may occur depending on the texture of the capsule, the needle point created, and the surgeon's actions. It is important to identify which pattern has been initiated almost immediately (within the first millimeter) after the puncture and initial drag of the needle have been made.

1. Single Incision Pattern

If the bent portion is fairly small, a single incision may develop. A curved incision is scribed by the needle to delineate the first 90 degrees of the capsulotomy (Figures 5–12, 5–13). A flap is turned, and this line is continued by drawing this flap around in a circular fashion (Figure 5–14). If a pupil parallel pattern is not observed at every moment, a straight line will start to develop. This may result in a capsulotomy that is too small or one that gets lost among the zonular fibers in the equatorial zone.

These can be saved sometimes, (Figure 5–15) but it is better at other times to reinitiate a capsulorhexis at the starting point going the other way or to

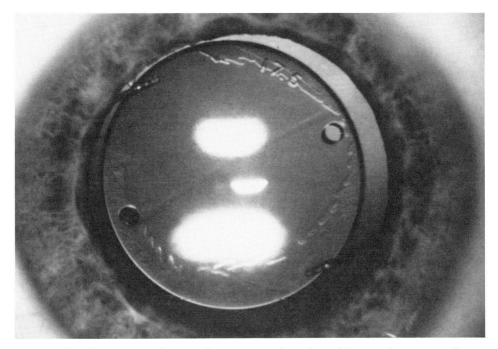

Figure 5–1. The size of an anterior capsulotomy seems just about right just after surgery. Note the anterior radial capsular tear at the 11 o'clock position.

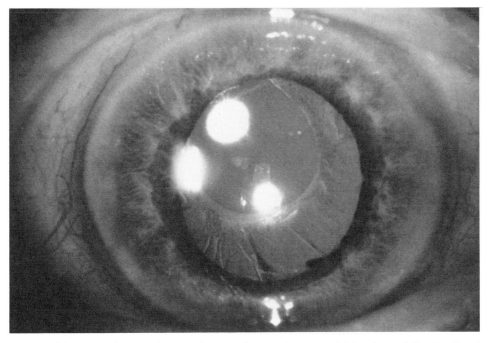

Figure 5–2. After several years, the anterior capsular opening may shrink substantially, drawing the opening toward an eccentric anterior capsular defect.

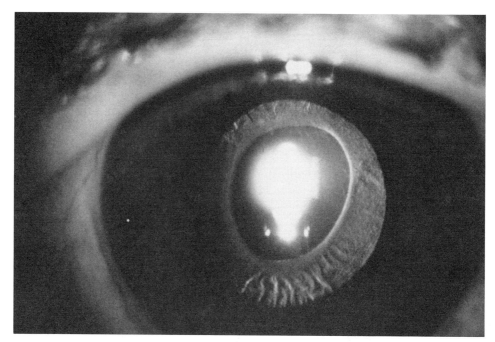

Figure 5–3. The anterior capsular opening was larger than this when originally created. As capsular fibrosis progresses, a sphincter–like effect seems to be present in some of the anterior capsular remnants created by capsulorhexis.

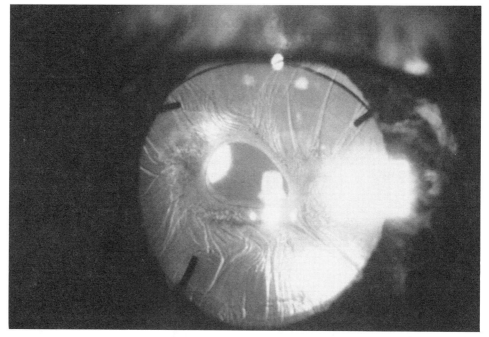

Figure 5–4. The extremely loose zonular attachments in this eye with pseudoexfoliation do not counteract the normal forces of progressive capsular fibrosis. The unopposed forces excessively contract the capsule, converting a 6.5–mm anterior capsulotomy into one that is hardly 2×3 mm. Obvious pseudophacodonesis is present as the 7–mm IOL—with curled–over haptics—wobbles with every eye motion.

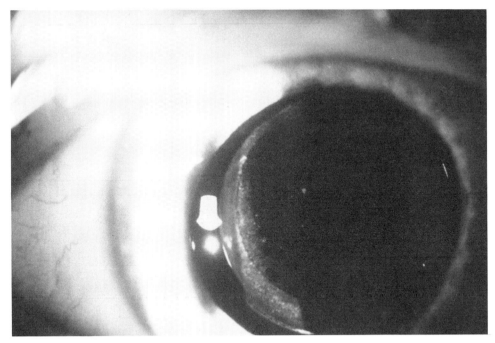

Figure 5–5. Proteinaceous debris is present on the surface of some IOL's after capsulorhexis. This reduces the functional visual surface of the IOL and is likely a product of inflammatory and lens epithelial cell interaction.

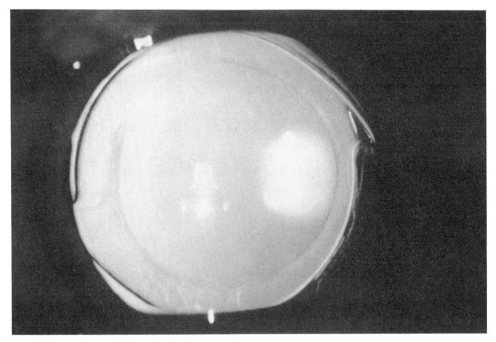

Figure 5–6. Some of the effects of the peripheral optical zone in this 7–mm multifocal lens will be unavailable for processing because of the shrinkage of the anterior capsular opening that has occurred with capsular fibrosis.

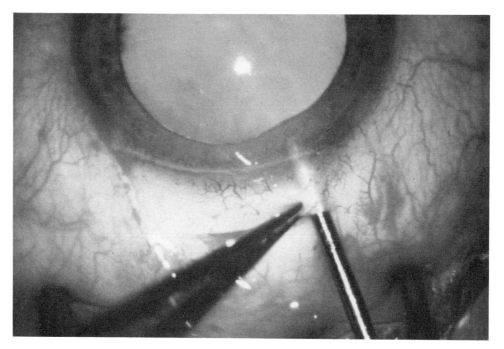

Figure 5–7. A 22–gauge blunt cannula is gently introduced while supporting the corneal edge to prevent trauma to Descemet's membrane.

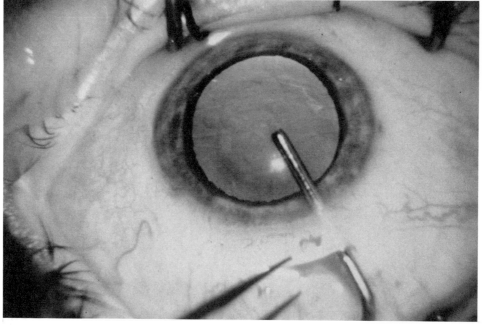

Figure 5–8. After it's obvious that the cannula has cleared Descemet's membrane, viscoelastic such as a Healon can be injected into the inferior anterior chamber. This forces the aqueous from the wound entry site while filling the anterior chamber from inferior to superior.

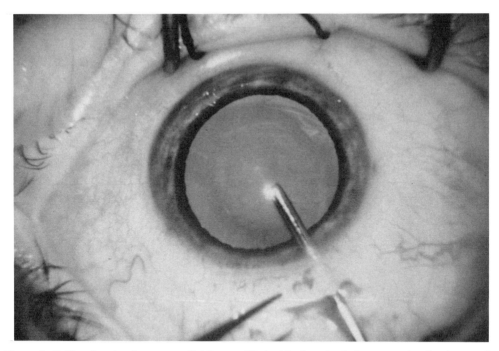

Figure 5–9. The chamber is ever so slightly overfilled with viscoelastic in preparation for the anterior capsulotomy. This overfilling is not extreme but just enough to make the individual rivulets of Healon merge and the pupil slightly expand.

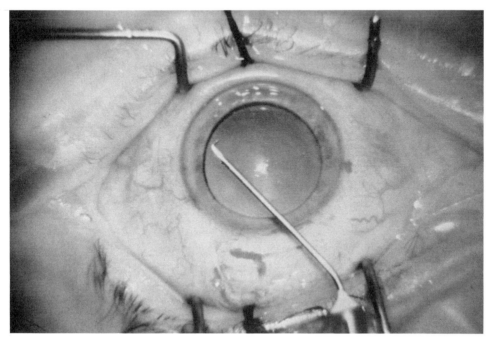

Figure 5–10. A 25–gauge needle has been bent in two places in preparation for creation of the anterior capsulotomy. The bend at the tip is made at a very sharp angle at a right angle and a 90–degree angle with a sharp corner to facilitate manipulation of the anterior capsular flap. A gentle bend is created at the base so that the handle may be held like a pencil.

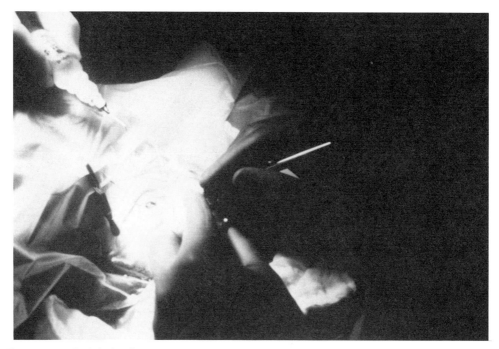

Figure 5–11. Both index fingers are used to guide the handle when creating the important anterior capsulotomy.

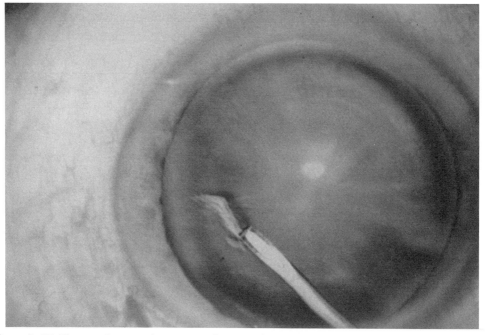

Figure 5–12. The anterior capsulotomy may be started as a puncture and become a single incision drawn with the needle. No flap of anterior capsule is seen at the initial stages of the anterior capsulotomy.

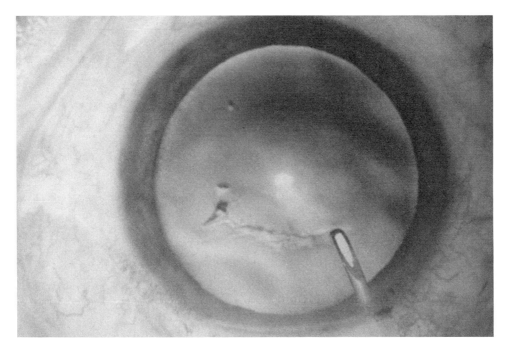

Figure 5–13. The line is drawn in curved fashion parallel to the pupillary border.

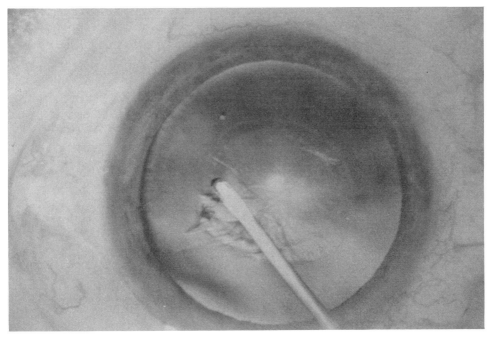

Figure 5–14. An anterior flap is created by lifting the central remnant slightly and the remnant dragged around in pupil parallel fashion.

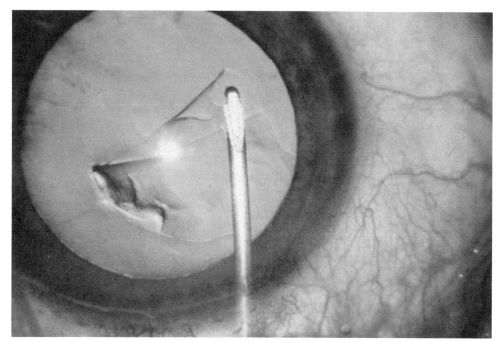

Figure 5–15. The anterior capsulotomy did not parallel the pupil for very long and extended out to the 10 o'clock position into the anterior equatorial zone. Some of the anterior zonular insertions were involved. In fact, zonular traction can be seen at the rhexis just at the 10 o'clock position. No anterior radial tear is present, but the capsular bag has been significantly structurally changed by the excursion into the equatorial zone.

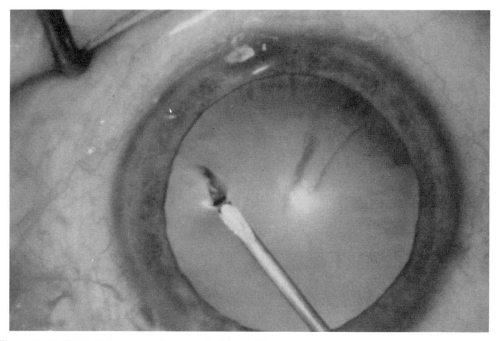

Figure 5–16. The initial puncture is created with a fairly relatively long portion of the needle.

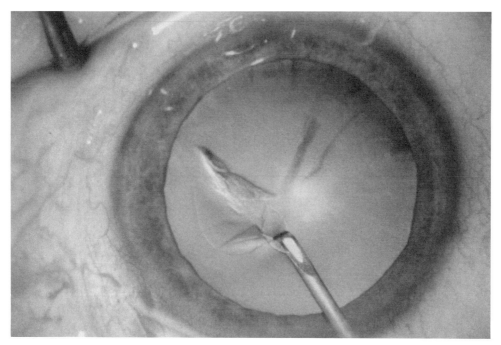

Figure 5–17. The flap is dragged so that the exterior tear forms the anterior capsulotomy with the very small interior tear basically pivoting around itself.

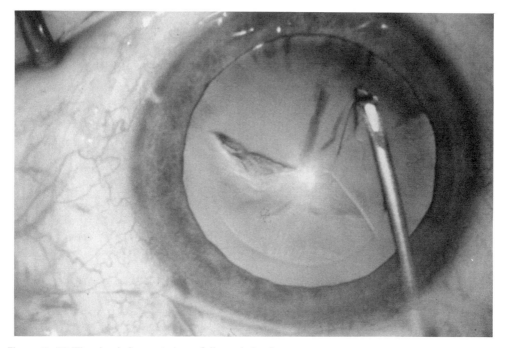

Figure 5–18. The rhexis is created carefully and slowly.

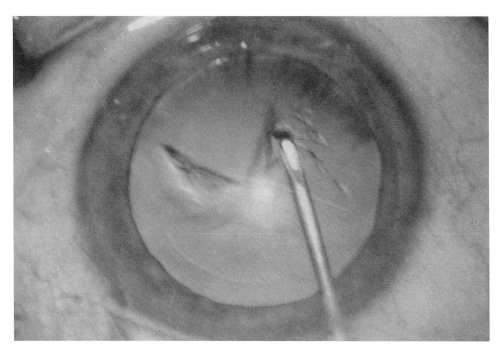

Figure 5–19. The final quadrant is involved with the capsulorhexis process. Frequent repositioning of the needle tip so that it is never more than 2 mm from the active peripheral tear edge is important or the capsulotomy will likely extend further peripherally than desired. See Figure 5–12. The final tear edge is prepared so that it is guided to be peripheral to the position of the initiation point.

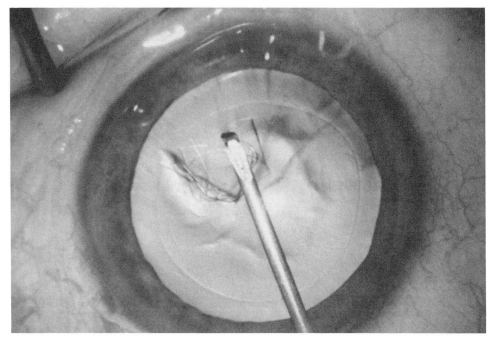

Figure 5–20. Torn and trailing capsular tissue is drawn centrally from time to time so that it does not interfere with the peripheral tear process.

Figure 5–21. The important capsular finish occurs as the end of the capsulotomy comes around just peripheral to the starting point. This includes the starting point in the central tissue that is removed.

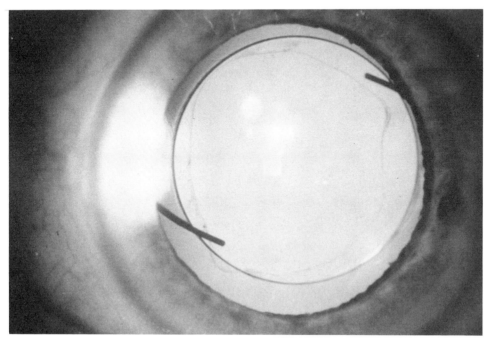

Figure 5–22. A wavering of the capsular edge is apparent from capsulorhexis tear "oversteer." But, on the whole, this capsulotomy looks structurally sound and is containing the lens well.

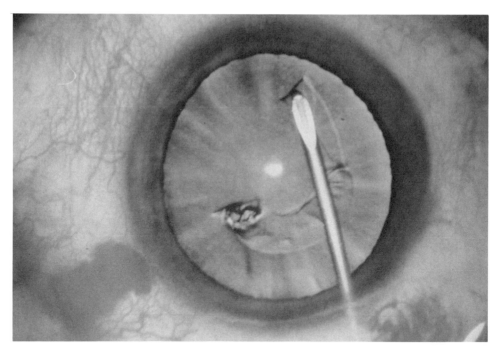

Figure 5–23. Closing angle tapering capsulorhexis flap pattern is seen.

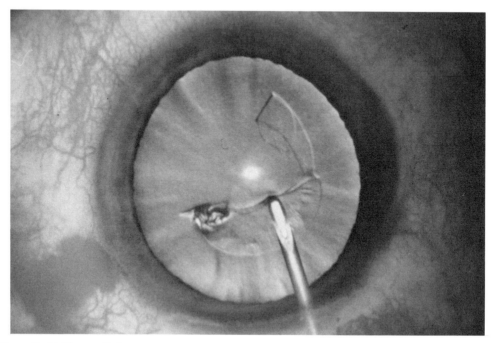

Figure 5–24. The needle has gently lifted the central and midperipheral portions of the anterior capsular flap.

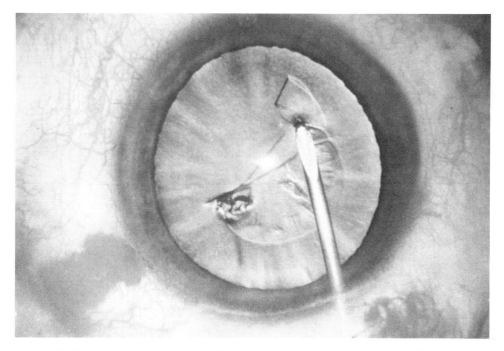

Figure 5–25. The peripheral tear line is continued without change, having recruited the entire central flap in the dragging process. This could have been done regardless of whether the initial tapering flap had fallen away.

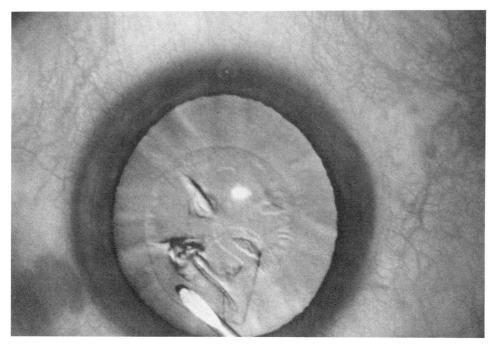

Figure 5–26. The final tear edge is guided so that it will be just peripheral to the starting point.

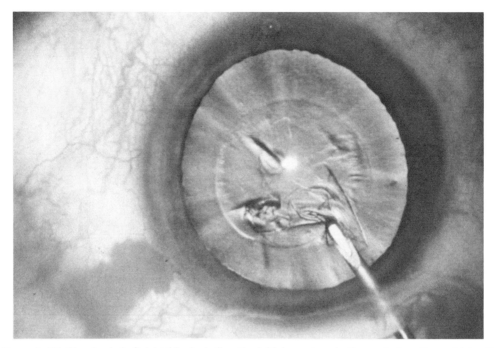

Figure 5-27. The rhexis is finished by including the initial puncture in the central section, which will be removed.

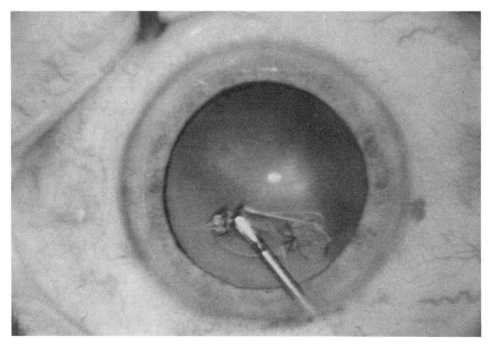

Figure 5-28. The central initiation point is connected by the central flap to the active peripheral tear site in the second quadrant. This radius connection is too small and will not permit the peripheral tear to develop in pupil parallel fashion. Several small tears are created central to the initiation point, heading toward the center of the rhexis remnant.

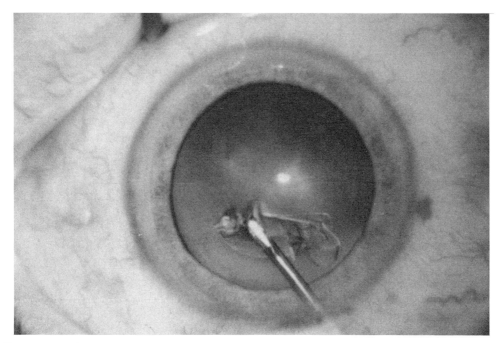

Figure 5-29. The second or third relaxing incisions have been made.

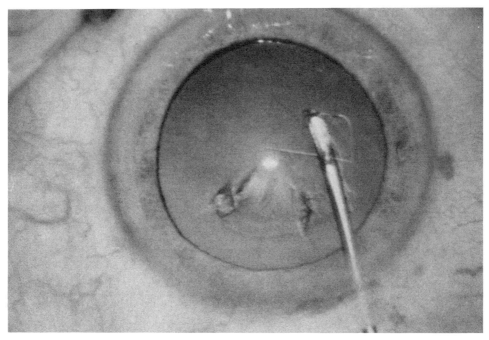

Figure 5-30. The peripheral tear can now be continued in pupil parallel fashion.

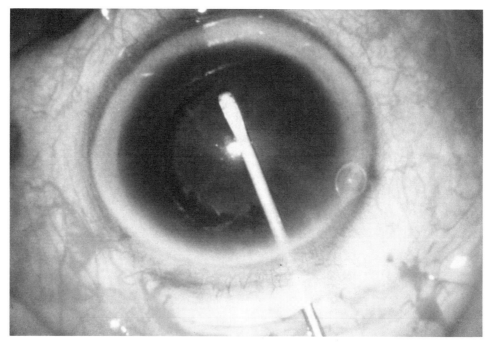

Figure 5–31. A very dense firm cataract in a patient with a fairly small eye and prominent arcus senilis is a good case in which to employ a multiple tear anterior capsulotomy.

just bring a can-opener pattern around to the weak spot from the starting point. It is important to meet or come around peripheral to the capsulotomy starting point at the conclusion of the capsulotomy. If this is done, the starting point will be included in the portion of the central anterior capsule that is to be removed. There will be no visible starting or end point in the remaining anterior capsular remnant. This is especially important if the single–incision pattern emerges just after capsulotomy, because the initiation point and initial capsulotomy will be central to that which is ultimately desired.

2. Opening Angle Flap Pattern

This pattern is more likely to be seen if the bent portion of the needle is slightly larger and impaled deeper so that each of its edges can start a tear so that a flap will result (Figure 5–16). This is the way I can best create consistent anterior capsulotomies and, for that reason, I prefer it over the other techniques discussed. I look for the flap to develop immediately after the first millimeter of dragging the needle. If present, the flap is carefully folded over, engaged and drawn by its apex so that the outer tear parallels the pupil (Figures 5–17, 5–18).

The tip of the needle should never be more than 2 mm from the tear site. Otherwise, the control, usually seen as a tendency to tear peripherally, will be lost (Figure 5–19). This becomes most critical at the last quarter of the capsulotomy when the finishing edge should be just peripheral to the starting edge. Excess capsule must be gathered up so that it does not interfere with the capsulotomy (Figure 5–20). The capsulotomy will finish just peripheral to the starting point (Figure 5–21).

"Oversteer" may be evident in the surgeon's initial capsulotomies (Figure 5–22), but this is better than losing control into the zonules and beyond or making the capsulotomy too small and ending up with anterior radial tears to the equator.

3. Closing Angle Flap Pattern

After the initial puncture, the flap apex is dragged parallel to the pupil, but the flap of anterior capsular tissue gets smaller instead of larger as the inner and outer tears converge (Figure 5–23). Either before or after this flap separates, the central flap of tissue is elevated first centrally, then peripherally, and a new outer tear is begun (Figure 5–24, 5–25). It is continued while keeping the flap away from the active tear site and finished just peripheral to the starting point (Figure 5–26, 5–27).

4. The Short Central Radius Flap

Sometimes a single incision will be initiated and turned through the first 90 degrees, but the central anterior capsular remnant radius is just too short to permit the peripheral tear from developing in pupil parallel fashion in the second quarter (Figure 5–28). This will result in a capsulotomy that turns too central in this second quarter and will ultimately be too small, making nucleus removal difficult. A relaxing incision or two are made near the pivot point of the central capsular attachment (Figure 5–29). This allows a central-peripheral tear flap configuration so that the peripheral tear can be pupil parallel (Figure 5–30). Actually, I now routinely make these relaxing incisions in radial fashion to the capsular center after the initiating flap tear. This establishes a perfect radial pivot for the creation of the outer tear every time.

Capsulorhexis in Children

Capsulorhexis in children is difficult for several reasons. First, pediatric anterior capsules are tough and tougher to tear well. Second, there is a tremendous elasticity to pediatric tissues. The scleras are spongy and the lenses soft. These features contribute to a relatively increased vitreous pressure situation, and the lens nucleus wants to come forward and actually express its gelatinous self while the anterior capsulotomy is being made. These tendencies make the capsulotomy extend further peripheral than desired and make it difficult to control all the way around.

One must make sure to start in an exaggerated central position, perhaps imagining a final goal of a 3.5 mm to 4.0 mm anterior capsulotomy. The capsulotomy will spiral out to at least 5.5 mm all on its own, and the finish will easily erase the starting point.

Capsulorhexis is an excellent technique, but it is not the best for all situations. At times, when visualization is extremely difficult or a minimal lift technique is necessary, a can-opener capsulotomy gives the most for the least risk (Figure 5–31).

Section III

PREPARATION FOR EMULSIFICATION

6

Managing the Small Pupil

William Fishkind,MD
Paul S. Koch, MD

Introduction

A small pupil in phacoemulsification is a pupil that is small enough to impede removal of the cataract. In most cases, it is between 2 mm to 5 mm in diameter, but in some eyes and for some surgeons, even a 7 mm pupil can be considered small. The absolute size of the pupil is just one of the relative factors in a surgeon's determination that a cataract will be easy to remove.

Other factors that contribute to a surgeon's decision to change the size of the pupil include:

(a) The hardness of the lens material relative to the size of the pupil.

(b) The status of the zonular support meshwork. For example, patients older than 90 and patients with pseudoexfoliation have weak zonules.

(c) The status of the anterior chamber. If the anterior chamber is shallow, it is more difficult to maneuver the phacoemulsification tip in the anterior chamber, and a larger pupil is helpful.

(d) The status of the cornea. Patients with Fuchs' Dystrophy and patients who are status post penetrating keratoplasty benefit from larger pupils, which make the phacoemulsification procedure easier.

(e) The performance of associated surgery. Surgery combined with trabeculectomy, for example, is a natural situation for a surgical expansion of the pupil that may, if the surgeon desires, be partially or completely closed at the end of the case.

Non-Surgical Treatment of the Small Pupil

If only a little enlargement is necessary, some viscoelastic can be used to enlarge the pupil during the capsulotomy.

If that is insufficient, the pupil can be stretched by using iris hooks placed through the surgical wound or one hook placed through the surgical wound and another through a side-port incision 90 degrees away.

Surgical Treatment of the Small Pupil

If the pupil remains too small after these measures, or if it is clearly necessary to obtain a significant degree of enlargement, some type of surgical intervention will be necessary. This can be in the form of a sphincterotomy pupilloplasty, a mid-peripheral (partial sector) iridectomy, a full sector iridectomy, or a keyhole iridectomy.

The Sphincterotomy Pupilloplasty

This technique helps the pupil to dilate while preserving its roundness and ability to dilate and constrict. It is discussed and described in Chapter 7 by Dr. I. Howard Fine.

The Mid-Peripheral (Partial Sector) Iridectomy

The iris is grasped with toothed forceps approximately one-third or one-half the distance from the chamber angle to the pupil. A full–thickness, mid-stromal iridectomy is then performed. Viscoelastic is placed above and below the iridectomy. Vanness scissors are then used to enlarge the iridectomy to the pupillary aperture, creating a partial sector iridectomy (Figures 6–1, 6–2).

This opens the pupil nicely and can be employed for moderately small pupils. It can be performed at twelve o'clock to provide access to the superior pole of the nucleus during two-handed phacoemulsification. A similar procedure can be performed at six o'clock by simply cutting the pupillary sphincter and iris to a point halfway between the pupillary margin and the chamber angle. This will provide access to the inferior pole of the nucleus for one-handed phacoemulsification (Figure 6–3).

The benefits of this technique include not having iris pillars to be caught in the phaco tip, as occurs with a keyhole iridectomy, and the preservation of a ring of peripheral iris to protect the zonules from the trauma of the moving phacoemulsification tip sleeve.

Sector Iridectomy

A more complete and permanent opening in the superior iris is through a complete sector iridectomy. Toothed forceps are used to grasp the iris near the pupil and then are withdrawn from the eye, bringing with them a large segment of the superior iris. The pupillary fringe is visible and, when it clears the incision, the entire piece of iris is excised with scissors (Figures 6–4, 6–5, and 6–6).

The full sector iridectomy makes a very large opening through removal of a significant amount of tissue. It cannot be repaired at the end of the operation.

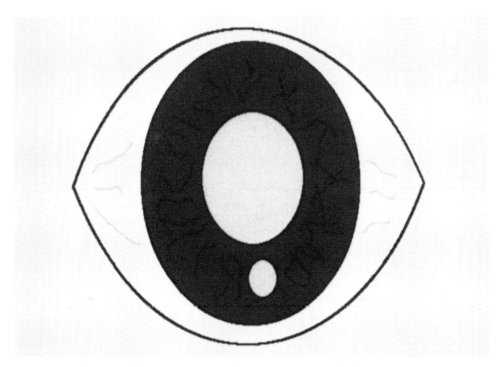

Figure 6–1. Partial sector iridectomy begins with a mid–stromal, full–thickness iridectomy.

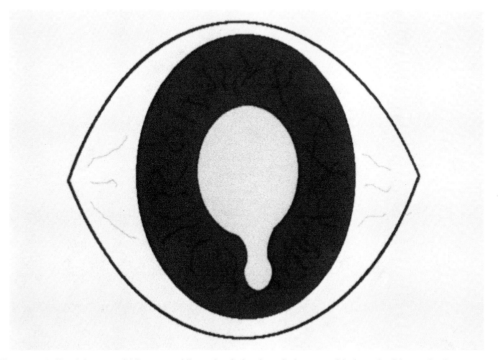

Figure 6–2. Partial sector iridectomy. Viscoelastic is placed above and below the iris, and scissors are used to enlarge the iridectomy to the pupillary aperture. Additional viscoelastic will enlarge the pupil even further.

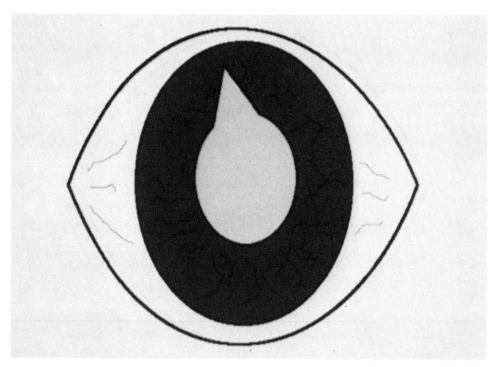

Figure 6-3. A partial sector iridectomy can be performed inferiorly simply by cutting the pupillary sphincter and iris to a point halfway between the pupillary margin and the chamber angle.

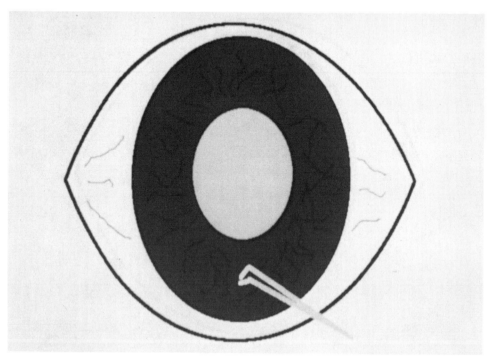

Figure 6-4. Sector iridectomy removed more tissue than the partial sector iridectomy. Forceps are used to grasp the iris near the pupil.

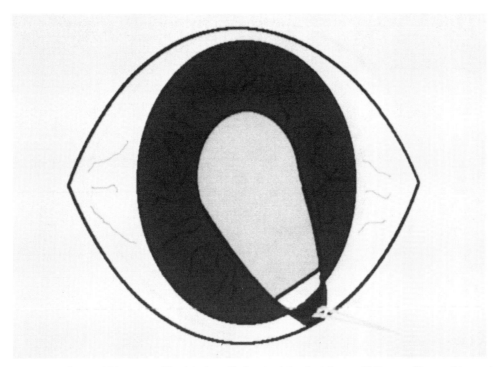

Figure 6–5. Sector iridectomy. The iris is pulled out of the incision until the pupillary sphincter is exposed, and then excised.

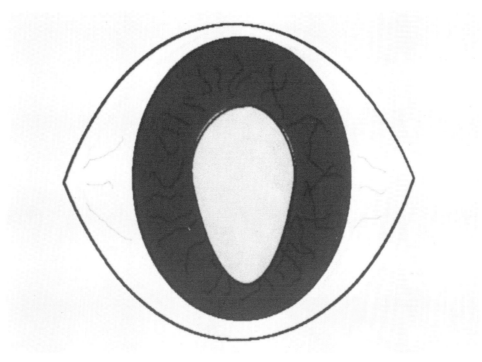

Figure 6–6. Sector iridectomy. A bridge of normal tissue can remain superiorly to assist in anterior chamber control. The rest of the iridectomy is removed and cannot be repaired afterwards.

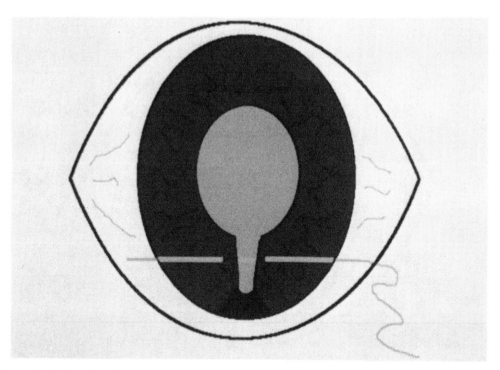

Figure 6–7. Post-placed sutures. The needle is passed through clear cornea, through each of the iris leaflets, and back out through the cornea on the other side.

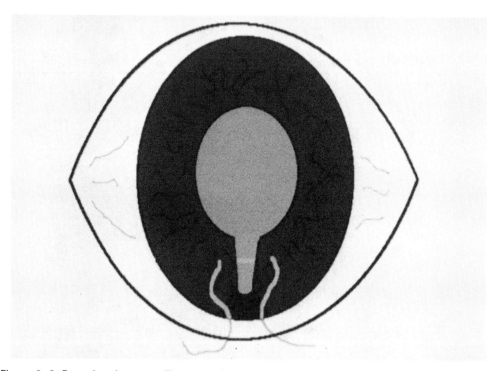

Figure 6–8. Post-placed sutures. The suture is cut from the needle and both ends are brought out through the surgical incision.

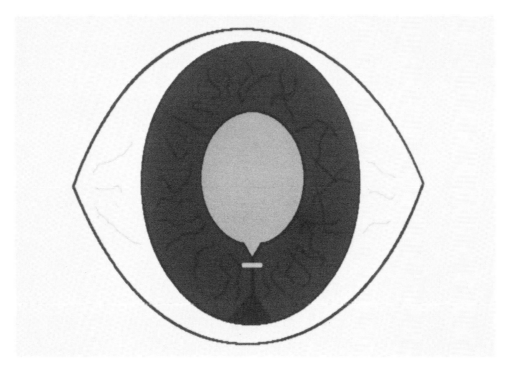

Figure 6–9. Post-placed sutures. The suture is tied, bringing the iris together. A slight gape is left at the pupil border to facilitate dilation and retinal visualization.

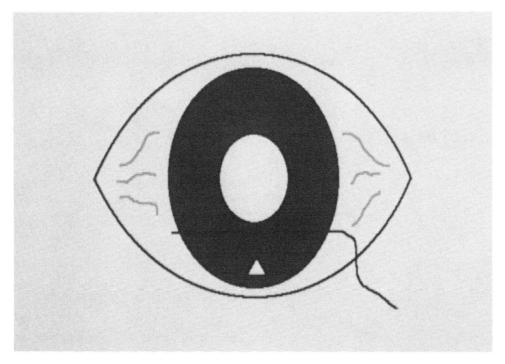

Figure 6–10. Pre-placed sutures. The needle is passed through clear cornea, through the iris on either side of the planned iridectomy, and out through clear cornea on the other side.

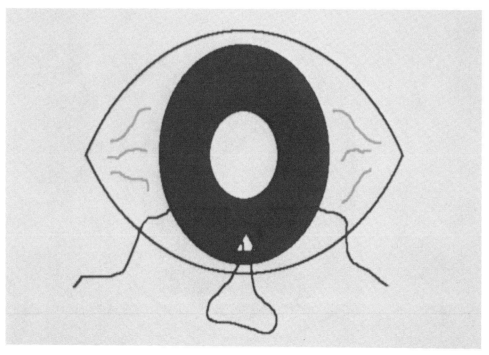

Figure 6–11. Pre-placed sutures. A hook is used to pull the suture from under the iris, through the peripheral iridectomy, and out the incision.

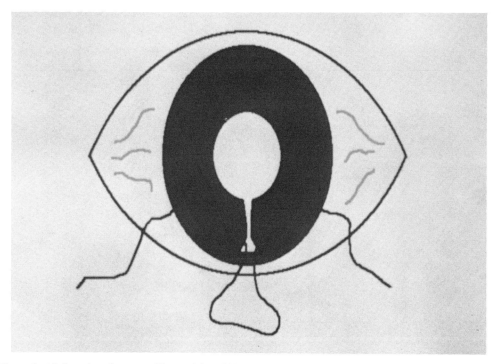

Figure 6–12. Pre-placed sutures. The peripheral iridectomy is converted to a keyhole iridectomy, taking care to avoid the suture.

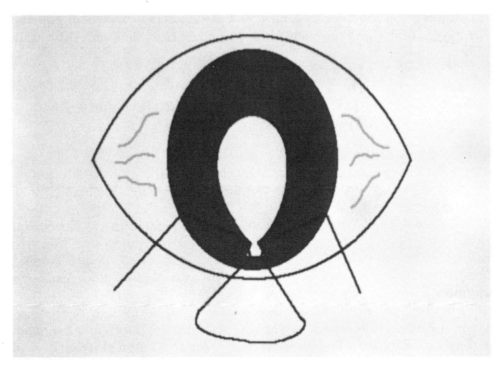

Figure 6–13. The suture is pulled tightly, taking out all of the slack. The pupillary edges of the iridectomy are drawn upwards by the tight sutures.

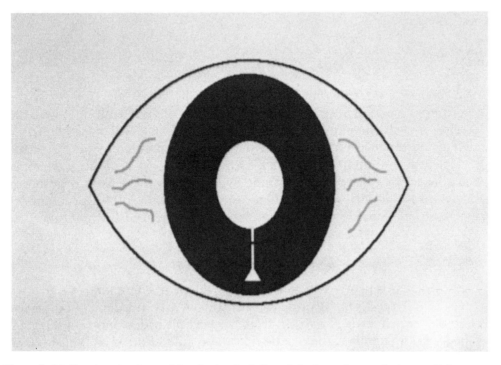

Figure 6–14. Pre-placed sutures. After the implantation of the lens, the needle is cut off the suture, both ends are pulled through the incision, and the suture is tied, approximating the iris.

This should be reserved for cases in which the aesthetics of a round pupil are unnecessary, or in those cases when the retina needs periodic and clear examination.

An alternative technique to make a sector iridectomy begins with a peripheral iridectomy. An iris hook is passed through the iridectomy and slips between the iris and the cataract. The pupil is engaged with the hook and is withdrawn from the eye, where it is excised.

Keyhole Iridectomy

As already noted above, this can be performed superiorly or inferiorly to provide access to the superior or inferior pole of the nucleus, depending upon the technique.

If performing the technique superiorly, the peripheral iris is grasped with toothed forceps and a full–thickness, peripheral iridectomy is performed. Viscoelastic is placed above and below the iridectomy, and the iridectomy is extended through the pupillary sphincter to create a full–thickness sector iridectomy.

If performed inferiorly, Vanness scissors can be placed into the anterior segment with generous amounts of viscoelastic, and the pupil can be cut from the pupillary margin to the chamber angle using the Vanness scissors.

The benefits of this procedure include a markedly enlarged pupillary aperture for much ease during phacoemulsification. There is a problem, however, created by the free pillars of the iris. They are flaccid and, therefore, are easily engaged by the phaco tip, causing shredding of the iris and a more difficult procedure.

Repair of Iridectomies

A surgeon may elect to repair or not repair the coloboma created by the mid-peripheral (partial sector) iridectomy or the keyhole iridectomy. If the iridectomy is performed at twelve o'clock, the upper lid will generally cover the defect and repair is not a necessity. However, an iridectomy performed at six o'clock will definitely need repair. The repair of a keyhole iridectomy prevents loose iris pillars from forming synechiae to the lens and posterior capsule. This is not only aesthetically pleasing, but more closely approximates normal anatomy. Sutures can be post-placed or pre-placed for iris repair.

Post-Placed Sutures

Once the lens implant is in position, the anterior chamber is filled with viscoelastic. A 10-0 Prolene suture on an Ethicon CIF needle can be passed through the cornea 90 degrees away from the keyhole iridectomy. Once passed into the anterior chamber, the needle tip can be guided through the proximal and distal iris leaves and then back out through the distal cornea. If the iridectomy is at twelve o'clock, both ends of the suture are gently teased through the twelve o'clock corneal scleral wound, and the suture is tied externally and slid down to the iris with minimal traction. The suture is cut close to the knot so there are no sharp ends to rub on the corneal endothelium (Figures 6–7, 6–8, and 6–9).

If the iridectomy is at six o'clock, a 1 mm stab incision is made directly inferior to the sector iridectomy, and the sutures are brought externally through this stab incision. The knot is also tied through this stab incision, and the sutures are cut. If the stab incision is iris parallel, it needs no further closure. If the stab incision is angled, it may need a single, buried 10-0 nylon suture for closure.

The suture should be placed approximately one-third of the way from the pupillary margin to the chamber angle. This creates a round–appearing pupil but provides a pupil large enough for a retinal evaluation and treatment if this becomes necessary at a later date.

Pre-Placed Sutures

After performing the peripheral or mid-peripheral iridectomy, viscoelastic is placed above and below the iris. A 10-0 Prolene suture on the CIF needle is then placed 90 degrees from the axis of the future keyhole or mid-peripheral (partial sector) iridectomy. The needle is passed through the iris on the pupillary side of the peripheral iridectomy, one-third of the way from the pupil to the chamber angle. The suture passes first through the future proximal iris leaf, then through the future distal iris leaf, and exits the anterior chamber and cornea after gentle pressure pushing it through the cornea (Figure 6–10).

A hook is then used to grasp the suture where it passed below the iris. This is done by placing the iris hook through the iridectomy and below the iris and gently grasping the suture and withdrawing it through the scleral wound. This loops the suture away from the future iridectomy so that Vanness scissors can then be placed through the peripheral iridectomy to the pupillary margin, cutting the pupil and creating a sector iridectomy (Figures 6–11, 6–12, and 6–13).

The entire phaco procedure is then performed. After implantation of the lens implant, the 10-0 nylon suture is first pulled taut, and the proximal and distal ends of the suture are brought out through the scleral wound, tied, and cut, as previously noted (Figure 6–14).

If performed at six o'clock, no iridectomy is necessary. The suture is passed through the iris one-third of the way from the pupil to the chamber angle. An iris hook reaches under the iris and pulls the suture loop into the pupil. A stab wound is then made inferiorly, and the loop of the suture is withdrawn through this stab wound. The iris is cut. After implantation of the lens, the suture is pulled taut, each end is withdrawn through the six o'clock scleral stab incision, tied, and cut, as previously noted.

Conclusion

The small pupil is one of the factors that causes even experienced phaco surgeons concern. It certainly is one of the key factors leading to intraoperative complications. A large pupil makes every case of phacoemulsification easier. These techniques, along with Dr. Fine's sphincterotomy pupilloplasty, provide a number of options for dealing with the small pupil and makes phacoemulsification in these cases easier and safer.

7

Pupilloplasty

I. Howard Fine, MD

The pupil that dilates poorly, is fibrosed, or hyalinized is frequently the determining factor in a decision not to proceed with phacoemulsification. The small pupil is most commonly managed by creating a sector iridectomy, which needs to be sutured in order to avoid edge effects from the implant and other potential sources of glare.

A sector iridectomy that has been sutured does not look natural and does not behave physiologically. The technique described below can be used in all cases, creating a pupil of adequate size for phacoemulsification, which will be cosmetically acceptable and physiologically functional postoperatively.

The Rappasso scissors (Figure 7–1) are shown compared to a standard U.S. dime. It has an outer cylinder with a blade at the end and a central rod with a blade at the tip that can be brought down against the blade attached to the cylinder, creating a shear force. The scissors are sufficiently small to fit through the smallest paracentesis incision used for capsulotomy.

The blades are made to shear by pressing the spring-loaded hemi-cylinders on the instrument's handle (Figure 7–2). The procedure is performed in an anterior chamber in which aqueous humor has been replaced by viscoelastic.

After lysing any synechiae with a cyclodialysis spatula, eight tiny sphincterotomies (Figure 7–3) measuring approximately 0.5 mm to 0.75 mm are cut at equal intervals around the pupillary border. This results in a dramatic increase in the size of the pupil (Figure 7–4).

It is important not to cut completely through the sphincter and into the iris stroma. The sphincterotomies need not be exactly radial, but they must interrupt the continuity of the sphincter edge, like a marginal myotomy in strabismus surgery. Almost all of the cuts appear radial following "healing".

After sphincterotomies have been cut in the inferior two-thirds of the pupillary circumference, the blades of the scissors will have to be rotated in order to make sphincterotomies in the superior one-third of the pupil. After completion of the sphincterotomies, the chamber is further deepened with viscoelastic, resulting in further dilation of the pupil, usually 5.5 mm to 6 mm in diameter. If, however, the pupillary size is still inadequate, a Lester hook

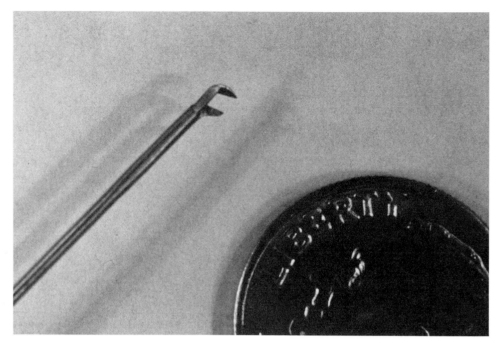

Figure 7–1. Blades are on the outer cylinder and central rod of the Rappasso scissors.

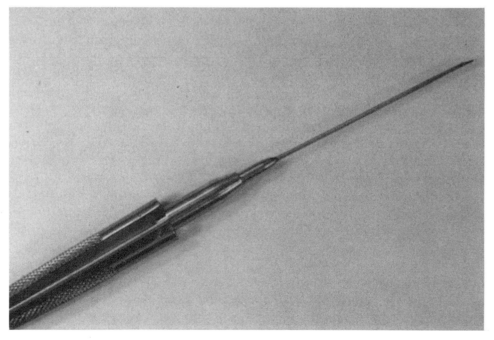

Figure 7–2. Spring-loaded hemi–cylinders control action.

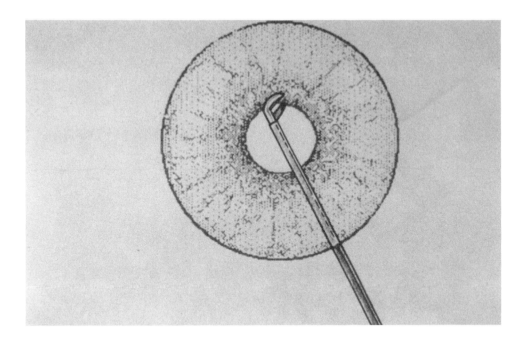

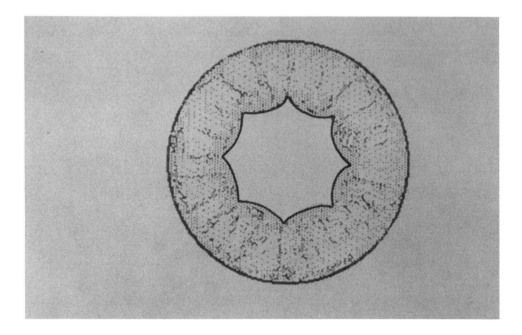

Figure 7–3, 7–4. Eight tiny sphincterotomies are cut at equal intervals around the pupillary border, which increases the pupil size.

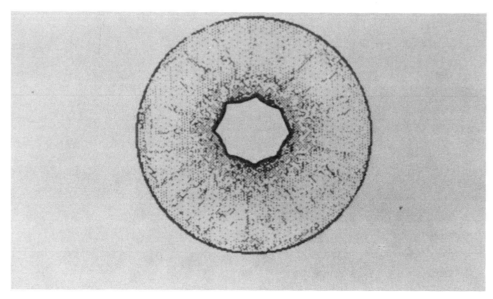

Figure 7-5. The pupil usually returns to a diameter very close to the preoperative diameter.

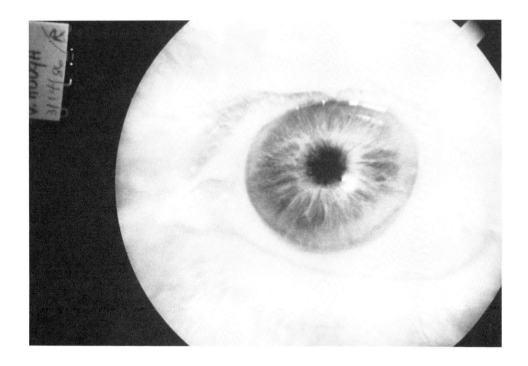

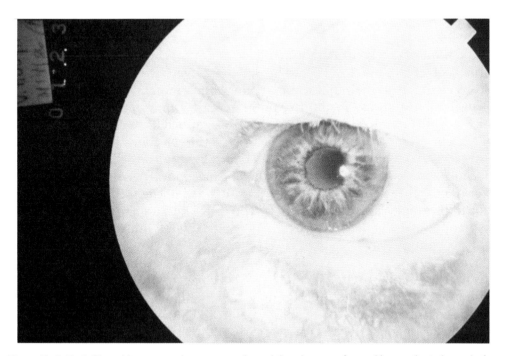

Figure 7–6, 7–7. Six sphincterotomies were made and the phaco performed in a patient shown before and after five minutes of dark adaptation.

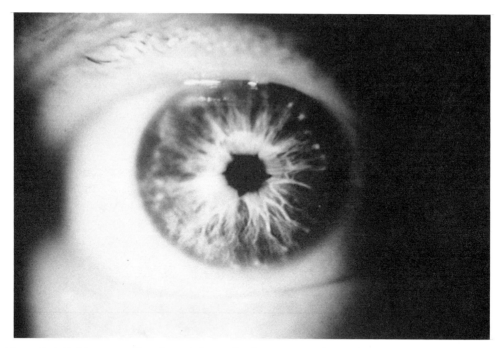

Figure 7–8. Sphincterotomies allowed phaco of a dense cataract in a patient who had been on miotic therapy for decades.

may be used to slowly stretch the pupil at each sphincterotomy site all the way to the root of the iris. I believe that this maneuver results in a rupturing of fibrous elements in the pupil while the muscular elements are only stretched. This will almost always result in a pupil at least 7 mm in diameter, through which cataract extraction can then be easily conducted.

Following completion of the cataract extraction and lens implantation, the Lester hook may be used gently to mechanically reduce the size of the pupil. Miochol (acetylcholine chloride, Iolab Pharmaceuticals) is instilled into the anterior chamber and Pilogel ointment is included in the medications instilled prior to patching. The pupil normally comes back to a diameter very close to the preoperative diameter (Figure 7–5), can be dilated much more easily, and reacts to light physiologically.

Figure 7–6 is a patient, postoperatively, who has had multiple episodes of iritis over a period of years, resulting in synechiae and fibrosis of the pupil in all positions except for the superior hour and one-half of the clock. Six sphincterotomies were made in the hyalinized, fibrosed and syneched portion of the pupil, after which the synechiae were lysed, the pupil was stretched, and phacoemulsification was performed.

The same patient (Figure 7–7) following five minutes of dark adaptation demonstrates physiologic dilation of the pupil. It is also of interest to note that although we are frequently assured that the upper lid covers a sector iridectomy, this would certainly not have been the case in this patient.

Figure 7–8 demonstrates the postoperative appearance of another patient whose pupil was hyalinized after decades of miotic therapy. Preoperatively, the pupil would not dilate to more than a 3–mm diameter. The procedure described above resulted in a pupil of adequate size for routine phacoemulsification of the dense cataract

8

Hydrodelamination & Hydrodissection

Paul S. Koch, MD

James A. Davison, MD

Terminology

Here's where some of the terminology gets confusing. A lot of different words have been coined that have the same meaning.

Hydrodissection generally means, as will be used in this chapter, the creation of a cleavage plane between the nucleus and the cortex. Fluid is injected between the nucleus and the anterior capsule, and the cleavage plane forms (Figures 8–1, 8–2). Note that the cleavage plane is not between the cortex and the capsule; if it were, no cortical aspiration would be necessary at the end of the case.

However, watch for other uses of this term. Dr. Fine likes to use hydrodissection to mean the creation of a cleavage plane between the inner nucleus and the outer nucleus. We prefer that this step be called hydrodelamination, hydrodelineation, or hydrodemarkation (Figure 8–3).

Hydrodissection

The incision is enlarged with the 3.0 mm keratome (Figure 8–4), and the 22–gauge cannula is reintroduced while still on the viscoelastic syringe (Figure 8–5). Approximately half of the viscoelastic material is removed for later use during IOL insertion. Remember, the anterior chamber had been overfilled to accomplish the anterior capsulotomy. Removal of some of the viscoelastic helps prevent iris prolapse during hydrodissection.

Balanced salt solution is injected under the anterior capsular remnant flap as superiorly as possible under the left and right (Figure 8–6) and inferior quarters. A curved cannula can be used to dissect at 11 o'clock, but this is rarely necessary (Figure 8–7). Many times, one of these injections will dissect posteriorly, but sometimes posterior dissection may not occur. Posterior hydrodissection is not necessary at this point for these methods, but peripheral equatorial dissection is an important start.

Hydrodelamination

This is also called hydrodelineation and hydrodemarkation, but for this discussion we will use hydrodelamination.

A small–gauge cannula, usually 26– to 30–gauge is attached to a 3–cc. syringe filled with BSS. It is placed in the eye and then in the nucleus at about the cut edge of the anterior capsule. It is passed into the nucleus until it meets resistance.

The point of resistance is where the soft outer nucleus ends and the firm inner nucleus begins. In some cataracts this is an obvious endpoint; in others, not so obvious. In some cases it's the point between the fetal nucleus and the adult nucleus. It's more noticeable in young cataracts and more difficult to find in the older, more firm ones.

At the point of resistance, the cannula is pulled back a fraction of a millimeter and the fluid is injected. The fluid passes into the body of the cataract and creates the cleavage plane, usually identified by the appearance of a golden ring around the inner nucleus. Sometimes only a dark separation plane is noticed.

If only a portion of the ring appears, it may be necessary to reintroduce the cannula in a different place and try to inject fluid again. The goal is a full separation of the inner from the outer nucleus.

In a very hard cataract, the inner nucleus might extend right out to the capsule. The cleavage plane may never be identified in a case like this, but the injection of fluid into the nucleus will help to separate it into its lamellae and make phacoemulsification easier.

In a very soft cataract, it may be possible to isolate several cleavage planes. Delaminating the cataract this way makes removal very easy because it is necessary to remove only a thin lamella of tissue at a time, sequentially removing the innermost nucleus and, then, progressively outer nuclear zones.

Hydrodissection is performed for the sake of followability. It separates the nucleus from the capsular bag so that it will be free to rotate within the bag, allowing each sequential piece of the nucleus to be rotated into the best position for removal, where some of the nucleus can be aspirated into the phaco tip. This would not be possible if the nucleus was still attached to the capsular bag through the cortical adhesions.

Hydrodelamination is performed for safety. The inner nucleus is the firm structure that causes all of the problems in phacoemulsification. It requires the most energy to remove, it's the hardest to emulsify, its edges are sharp and can cut the capsule, and if brought up through the capsulotomy too soon, it can split the capsulotomy edge, causing a rent to pass around toward the equator.

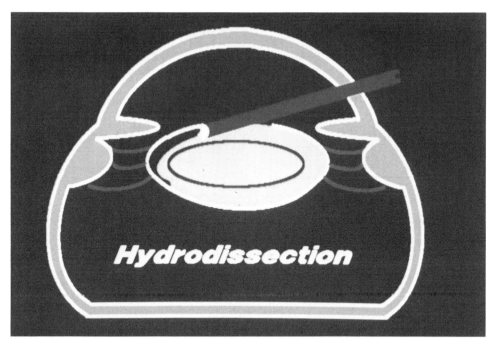

Figure 8–1. Hydrodissection creates a cleavage plane between the nucleus and the cortex. The cannula is placed just under the anterior capsule and fluid is gently injected.

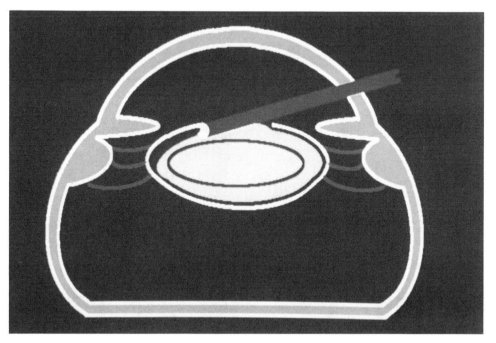

Figure 8–2. A fluid wave passes under the nucleus and completes the separation of the nucleus from the cortex.

Figure 8–3. Hydrodelamination (also called hydrodelineation or hydyrodemarkation) is performed by placing a 30–gauge cannula into the body of the nucleus until it meets resistance. Then fluid is injected, creating a cleavage plane between the inner nucleus and the outer nucleus.

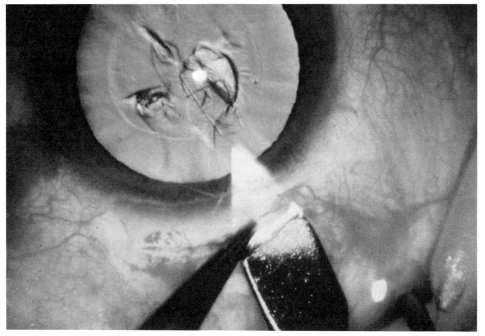

Figure 8–4. A keratome is used to enter the anterior chamber, making sure the keratome follows the incision created with the scleral pocket and the initial entry with the super blade. (Note: right hand scleral flap is too thin here. See Figure 22–18.)

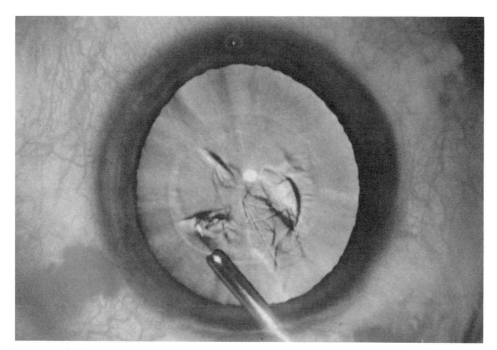

Figure 8–5. A 22–gauge cannula is placed on the viscoelastic syringe, and some of the viscoelastic is removed to prevent iris prolapse during hydrodissection. Care must be taken to remove sodium hyaluronate slowly or air will be drawn in around the hub. This technique is not as effective if Viscoat is used.

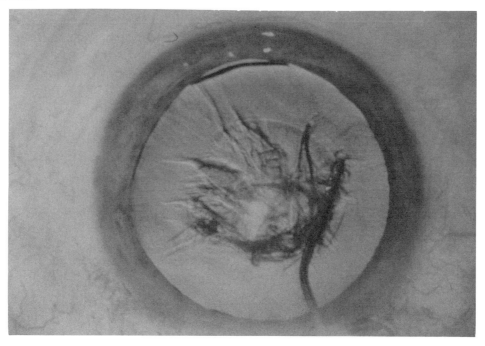

Figure 8–6. Slow gentle hydrodissection is begun with a 30–gauge cannula. Always watch carefully for early signs of iris prolapse.

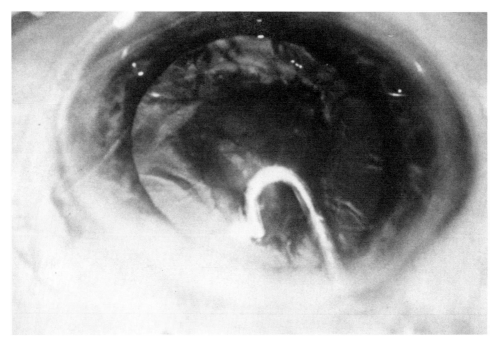

Figure 8–7. A curved cannula is an excellent device for separating the superior nucleus from cortex efficiently. It is particularly valuable in softer nuclei when subsequent manipulation will be more difficult. (Photograph courtesy of David Dillman, MD)

If it was possible to separate the inner nucleus from the softer nucleus, we would gain a significant margin of safety. The firm nucleus could be emulsified within a thick cushion of soft outer nucleus. It would be a lot like working within a layer of foam rubber. The capsular bag would be protected by this cushion, and unwanted movements of the firm inner nucleus would be less likely to cause capsular tears.

Separating the nucleus into an inner one and an outer one makes the job of phacoemulsification easier because it is necessary to attack only one piece at a time. It's a lot easier to devote one's attention to a 7 mm inner nucleus sitting in a bed of soft outer nuclear tissue than it is to work on a 9 mm nucleus sitting directly on the posterior capsule.

The operations that use hydrodelamination are grouped in the category of Endocapsular Phacoemulsification With Nuclear Cleavage. These include the Chip & Flip and Spring Surgery. Hydrodelamination is also a key step in a cracking procedure, Fractional 2/4 Phaco.

Section IV

PHACOEMULSIFICATION TECHNIQUES

9

Introduction to Phacoemulsification

James A. Davison, MD

Introduction

In the mid 1980's, relatively simultaneous presentations of what came to be know as capsulorhexis were given by Doctors Cal Fercho of North Dakota (Welsh Cataract Congress, 1986), Howard Gimbel of Canada (ASCRS film festival, 1985), John Graether of Iowa (Welch Cataract Congress, 1986), and Thomas Neuhann of West Germany (German Ophthalmology Society, 1985) (Figure 9–1).[1,2,3] Although in different locations around the world, these phacoemulsification veterans had all arrived at the same evolutionary point, realizing the advantages of phacoemulsification completely within the capsular bag. They also appreciated the benefits that an architecturally perfect capsular bag gave IOL fixation and centration (Figure 9–2). They were at this point because they could already emulsify a nucleus within the intact capsular bag using a can–opener anterior capsulotomy. Anterior radial capsular tears common to can–opener capsulotomies had frustrated them in their drive for anterior capsular remnant perfection. Various methods, including thermal capsule cutting cautery, (Figure 9–3) were tried to create a stronger anterior capsular edge that would resist the troublesome tear extensions. They needed to discover a method that would more consistently preserve an intact anterior capsular remnant after phacoemulsification, and they did. Surgeons who were practicing planned extracapsular methods had also realized the importance of an intact capsular bag, but nucleus expression necessarily disrupted the structural integrity capsular bag.[4,5,6] Until then, the capsular bag had been traditionally, if rather loosely, defined as whatever capsular structure remained for possible IOL fixation after any form of extracapsular cataract extraction. With the introduction of capsulorhexis, the definition was expanded and simultaneously refined to include the structurally symmetrical capsular structure that remained after partial anterior capsulectomy by continuous tear circular capsulotomy. The requisite emulsification process basically combined an inferior nuclear attack contained within the architecturally improved capsular bag and eventually has become known as "capsular bag phacoemulsification," what I consider a form of endophacoemulsification.

remained after partial anterior capsulectomy by continuous tear circular capsulotomy. The requisite emulsification process basically combined an inferior nuclear attack contained within the architecturally improved capsular bag and eventually has become known as "capsular bag phacoemulsification," what I consider a form of endophacoemulsification.

Even before this time, Drs. Hara and Hara were carrying out important pioneering work developing subcapsular[7,8] and intracapsular[9] phacoemulsification. An endocapsular surgical technique had been described in elegant photographic detail as well.[10] In most early endophacoemulsification techniques, the cataract was emulsified within a relatively intact lens capsule. Only a small entry perforation, a true capsulotomy, had been made in the most superior anterior capsule. An anterior capsulectomy was not performed. Many times the central anterior capsule was retained during emulsification and, depending on technique, left intact after IOL insertion. Pars plana techniques of endocapsular surgery had also been pioneered[11,12] but were not ever in substantial use by the cataract-IOL surgeon.

As of the late 1980's, the popular term "endophacoemulsification" not only defined a specific surgical technique through a small capsular puncture, but more generally through common usage served as a category designation for a number of similar phacoemulsification techniques. This category included forms of surgery accomplished under an anterior capsular flap after a small anterior capsulotomy (intercapsular phacoemulsification) and surgery within the capsular bag after a small anterior capsulectomy (capsular bag phacoemulsification). Ultimately, only micropuncture techniques should actually be called endophacoemulsification with the two current subcategories, intercapsular and capsular bag, becoming distinct categories themselves.

Three other location–specific phacoemulsification schools were already in existence as these various forms of endophacoemulsification emerged. Two of these had been firmly established as classical methods and one transitional school had evolved toward an endophacoemulsification discipline.

The anterior capsulotomy techniques of the two classical schools, anterior chamber phacoemulsification[13] and iris-plane phacoemulsification[14] had several things in common. Both techniques required a large anterior capsulectomy so that the nucleus could be removed, either partially or completely from the capsular bag, and be manipulated into a relatively safe place for exposure to the phacoemulsification tip. These large capsulectomies were created by the connection of a series of fairly large multiple peripheral capsulotomy tears (Figures 9–4 through 9–13). Any remaining anterior capsular flap tissue was usually viewed as an obstacle and inconvenience. An intact capsular bag was not important to either technique because the intraocular lens was placed in the anterior chamber in the first and was embraced by the ciliary sulcus in the second. In ciliary sulcus fixation, only the posterior capsule was needed to help stabilize the IOL and keep it from falling posterior. Remaining anterior tissue could actually contribute to asymmetric bag-sulcus haptic positioning and IOL decentration. The capsulectomies in either case might have afforded an intact bag with small anterior capsular flaps, but any hope of structural competency of the three–dimensional capsular bag was sacrificed during the nucleus relocation process.

Some surgeons who desired capsular bag fixation of intraocular lenses but had not yet learned of capsulorhexis were practicing phacoemulsification techniques in what I call the transitional school. The minimal-lift bimanual technique[15] and the more advanced roundel technique[16] (Figures 9–14 through 9–17) are good examples of transitional methods. The anterior capsulotomy techniques had several points in common. The resultant anterior capsulectomies were small at about a 5.5 mm diameter and were composed of relatively central very small multiple capsulotomy tears (Figures 9–18, 9–19, and 9–20). I call these transitional techniques because, while they attempted improvement over the classical methods, they were imperfect. The objectives of phacoemulsification entirely within the capsular bag were largely but not completely accomplished.[17] Capsular bag perfection was not uniformly achieved. Single or multiple anterior radial capsular tears developed frequently with either technique. They usually occurred superiorly in the minimal-lift method because this method still required an attack of the superior nuclear rim (Figures 9–21, 9–22, and 9–23). They occurred inferiorly in the roundel because this was one of the first methods to attempt inferior nuclear attack as an initiating component of nuclear rim debulking and removal (Figure 9–24). These capsular bag configurations led to fairly stable and relatively symmetrical capsular bag fixation (Figure 9–25), but a trend persisted toward minor, usually subclinical late optic decentration toward the anterior capsular defect (Figures 9–26, 9–27, and 9–28).

Phacoemulsification surgeons who did not practice within either of the classical techniques, one of the transitional variations, or by one of the new endo-phacoemulsification methods fell into one last category with a problem similar to their colleagues who practiced nucleus expression. This school might be called "mismatch phacoemulsification". Many of us unknowingly practiced its mismatch method. David Apple, M.D., and his colleagues have shown in their pioneering clinicopathologic studies that the most common problem in extracapsular cataract IOL surgery might result from the surgeon's mismatching one of the two classical large anterior capsulotomy phacoemulsification methods or a nucleus expression technique with the desire for capsular bag IOL fixation.[18,19,20,21] Many times the results of this mismatching were asymmetric bag-sulcus haptic placement and resultant optic decentration. IOL haptic knees were placed under small anterior capsular flap remnants, which necessarily remained after these techniques that required large anterior capsulectomies and nucleus relocation. The anterior capsular flap remnants could unfold, allowing the haptic knee to escape into the ciliary sulcus, or the unfolding would be incomplete and just allow the knee to migrate toward the sulcus while still peripherally contained by the partially unfolded flap (Figure 9–29). Some of these decentration problems were solved by the long C–shaped haptic IOL design of Bill Simcoe. The haptics were able to bridge the defects created in the equatorial zone of the lens capsule so that adjacent anterior capsular flaps might be recruited to support and center the IOL (Figures 9–30, 9–31).

As of late 1989, four major location–specific phacoemulsification schools are distinct: anterior chamber, iris-plane, transitional, and endophacoemulsification.

Subsequent to the capsulorhexis presentations, many different forms of endophacoemulsification evolved. These almost innumerable personal variations fall into one of the two aforementioned subcategories of endophacoemulsification, depending on the disposition of the anterior capsule.

I call the first subcategory of endophacoemulsification "capsular bag phacoemulsification."[22] In this technique, or more appropriately, in these phacoemulsification strategies, the final optical anterior capsulectomy is created by capsulorhexis before phacoemulsification is accomplished entirely within the capsular bag. The nucleus is removed by dividing it into smaller components. This removal involves a physical separation of nuclear fibers that compose the posterior central and or peripheral nuclear plate. This is accomplished with a folding over and fracture of the deep mid-peripheral nuclear plate by suction of the peripheral anterior nuclear remnant or with a bimanual separation of the central and peripheral nuclear plate into quarters or smaller pieces. The separation of nuclear fibers created by either method results in a fracture–like appearance in the full thickness of the deep nuclear plate. Doctors Howard Gimbel and John Shepherd were the first to introduce and popularize the bimanual posterior central nuclear separation techniques. Doctor Gimbel's popular "Divide and Conquer" endolenticular method is a good example of the surgical method (Figures 9–32 through 9–36). Doctor John Shepherd's in situ fracture surgery is another elegant derivation of this method (Figures 9–37, 9–38).

In the second endophacoemulsification subcategory, which I categorize as intercapsular phacoemulsification, a small superior anterior capsulotomy is created and the lens emulsified between the anterior and posterior capsules (Figure 9–39). The final optical anterior capsulectomy may[10,24,25] or may not[7,8,9,24] be created at some point after phacoemulsification (Figures 9–40, 9–41).

The main advantage to emulsification under the anterior capsule may be increased corneal endothelial cell protection.[25–29] However, intercapsular techniques, at least in my hands, seem to be relatively more difficult than two-handed surgical methods. Further, endothelial cell survival has not been shown to be superior to any of the capsular bag endophacoemulsification techniques after 5 mm central capsulorhexis. In its current evolutionary state, its two main apparent disadvantages are a relatively high posterior capsular rupture rate and frequent anterior radial tear extension into the capsular bag equatorial zone.[25,26]

Phacoemulsification within the capsular bag is the subject of Chapter 12. It is my preferred technique and it is, or rather they are, the methods that will be carefully analyzed herein. There are two advantages to phacoemulsification within the capsular bag that make it very efficient, very safe, and, thus, very appealing. First, there is relative physical containment of a safe bimanual emulsification process and, second, preservation of a relatively inert architec-

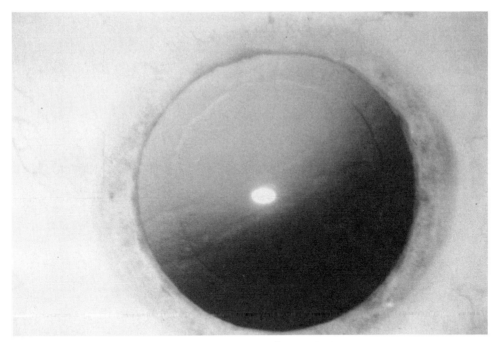

Figure 9-1. Capsulorhexis anterior capsulotomy as performed by Dr. Thomas Neuhann, West Germany. Note the perfect circular nature of the capsulorhexis opening. (Photograph courtesy of Dr. Thomas Neuhann)

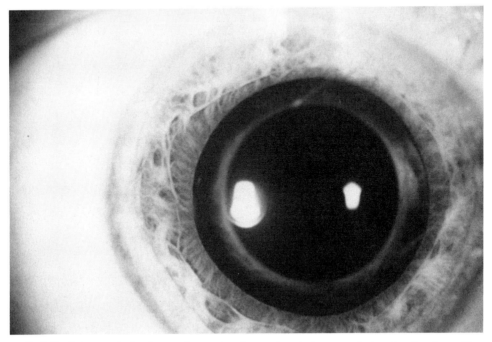

Figure 9-2. Perfect capsulorhexis opening embracing and centering a posterior chamber lens. (Photograph courtesy of John Graether, MD)

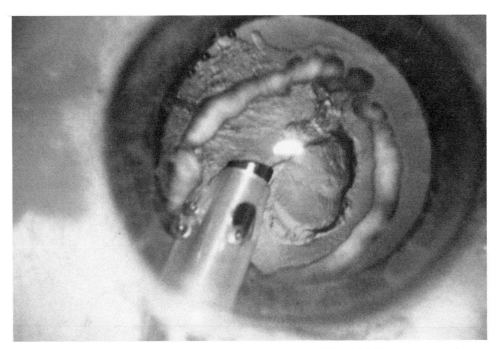

Figure 9–3. Capsule cutting cautery as designed by John Graether, MD, with MIRA corporation (Boston, Mass.) in 1980 was used to try to create a more perfect anterior capsular remnant.

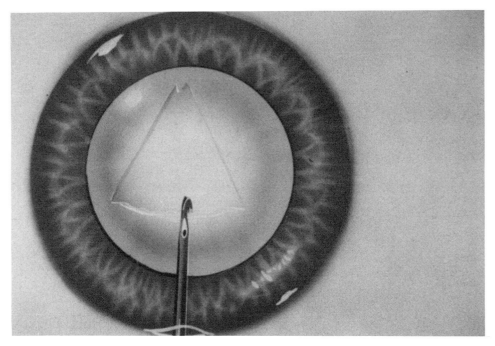

Figure 9–4. Large connected linear tears in the anterior capsule initiated Dr. Charles Kelman's anterior capsulotomy, which he uses for anterior chamber phacoemulsification. The even edge of the tears are similar to those seen in capsulorhexis except that they are not curved and do not parallel the pupil. (Photograph courtesy of Dr. Charles Kelman)

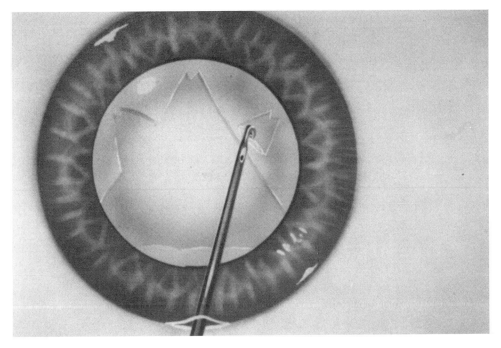

Figure 9-5. Smaller tears are used to expand the anterior capsular defect to make anterior chamber nuclear relocation possible. (Photograph courtesy of Dr. Charles Kelman)

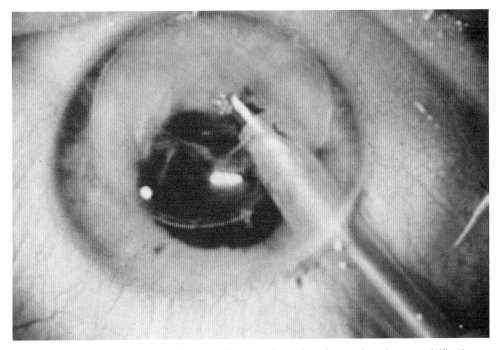

Figure 9-6. The large anterior capsulotomy allows nuclear relocation so that phacoemulsification can be efficiently accomplished in the anterior chamber. (Photograph courtesy of Dr. Charles Kelman)

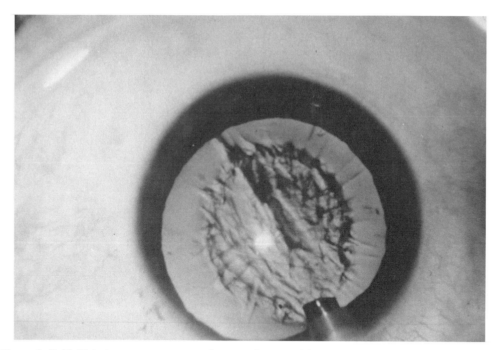

Figure 9–7. Multiple-tear can-opener anterior capsulotomy of 35 to 50 small tears. The capsulotomy in this figure is smaller than traditionally recommended, but the iris plane technique can still be demonstrated. It is slightly oval and will leave more generous anterior capsular flaps than ordinarily recommended.

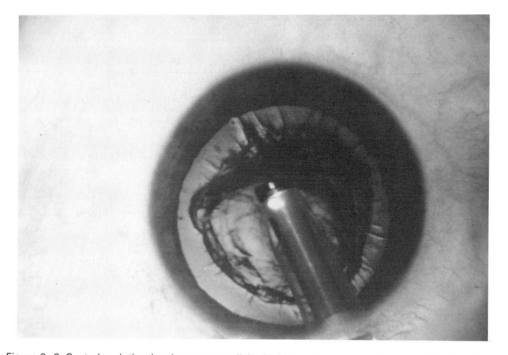

Figure 9–8. Central sculpting has been accomplished with the 15–degree phacoemulsification tip.

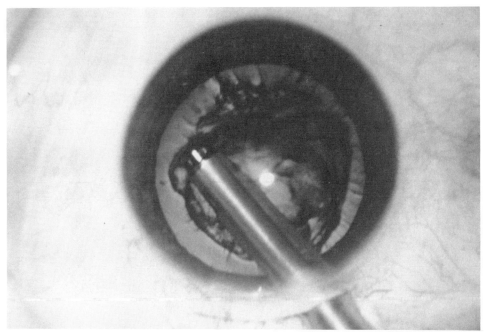

Figure 9-9. The practical sculpting limits of the 15-degree tip have been reached. Any further sculpting attempts may result in tip occlusion and penetration of the nuclear rim by the phacoemulsification tip.

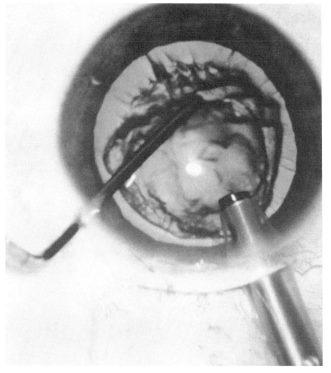

Figure 9-10. The phacoemulsification tip and the cyclodialysis spatula are in position to accomplish important superior pole prolapse.

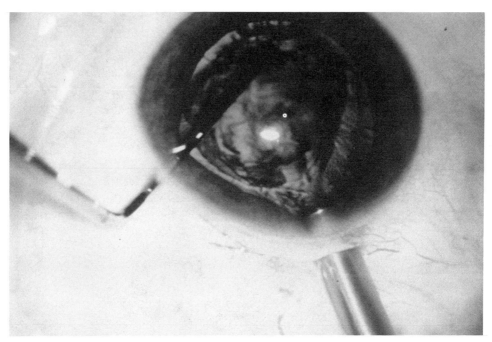

Figure 9–11. The superior pole prolapse maneuver is initiated by creating an increased relative vitreous pressure that pushes the superior pole of the nucleus anterior. This is accomplished by reducing the anterior chamber pressure and volume by going briefly into foot position O, thus cutting the infusion off. The superior pole of the nucleus comes forward because the inferior nucleus is trapped and held posterior and inferior by the cyclodialysis spatula.

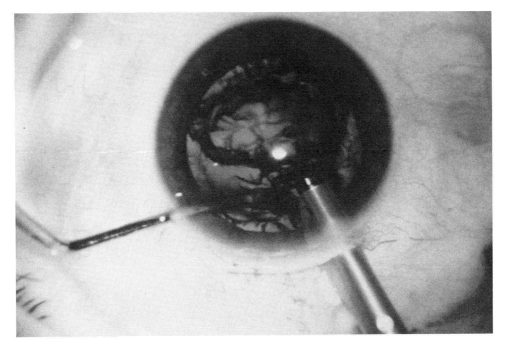

Figure 9–12. In the process of allowing the superior pole to come forward, an anterior radial capsular tear has been necessarily created. This tear allows the rest of the operation to proceed. Without it, the superior nuclear pole will not present anteriorly in the iris plane. Once it has come forward, the cyclodialysis spatula position is quickly changed so that it supports the superior nuclear pole in the iris plane. The infusion is turned on again by returning to foot position 1 so that the posterior capsule is pushed posterior and the anterior chamber is expanded. Once the surgeon makes sure that no anterior capsule is engaged in the tip (actually confirming that an anterior radial capsular tear has occurred), emulsification of the superior pole can be begin. Usually at least one other superior anterior radial tear has occurred as well.

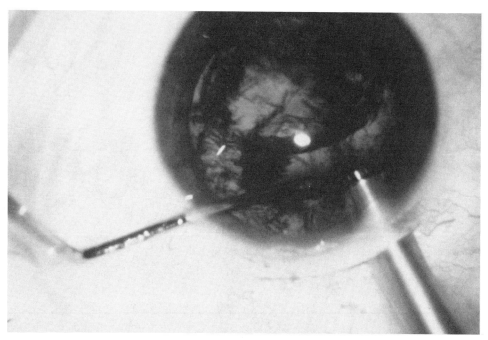

Figure 9–13. After the initial superior rim is emulsified, the nucleus is allowed to drop back. It is rotated by the cyclodialysis spatula so that roughly two or three new clock hours of nuclear rim are presented to the phaco tip access zone superiorly. The prolapse maneuver is repeated so that the superior pole will present itself just as it had initially. Emulsification proceeds with exposure of new rim tissue with each rotation.

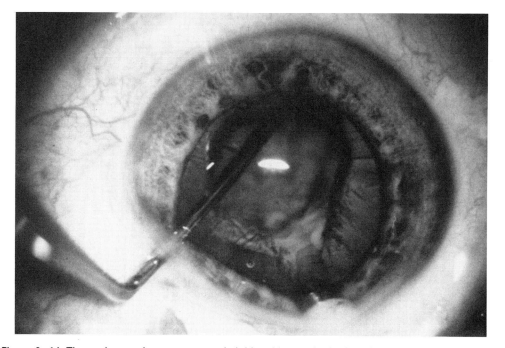

Figure 9–14. The nuclear prolapse maneuver is initiated just as in the iris plane technique.

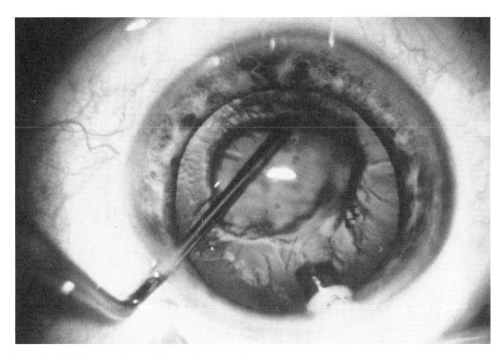

Figure 9–15. The superior nuclear pole comes forward, but it is not allowed to come as far forward as in the iris plane technique. This permits a more generous anterior capsular rim to remain and usually helps to insure that only one anterior radial defect is created.

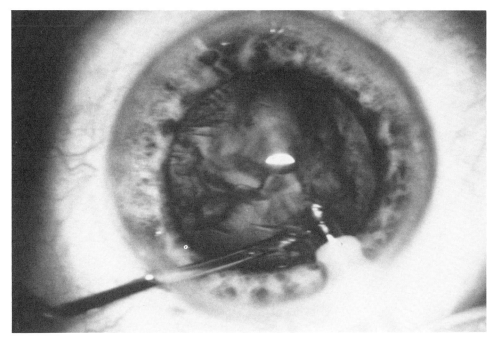

Figure 9–16. A new section of nuclear rim is exposed. Only the anterior and mid-level portions are attacked. The rim is not lifted high enough to gain to the deeper peripheral layer. This elevation of the superior nuclear rim is substantially less than that achieved in an iris-plane technique.

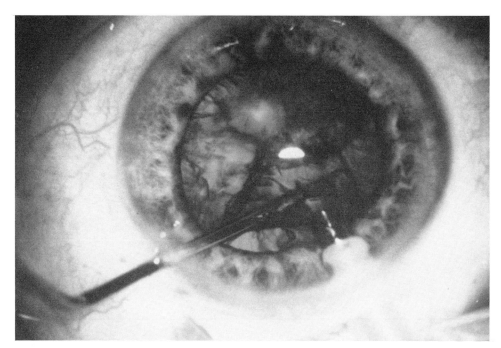

Figure 9–17. As further sections are rotated into position, exposure improves because the thinner nuclear rim 180 degrees away from the superior position has been thinned and debulked initially.

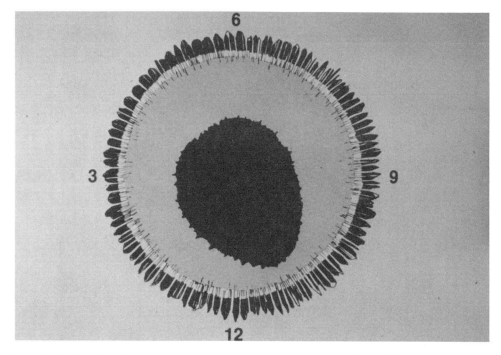

Figure 9–18. Drawing of an anterior capsulotomy tear pattern for a minimal-lift technique. Note its eccentricity, which facilitates superior rim exposure and anterior radial tear formation at 11 o'clock.

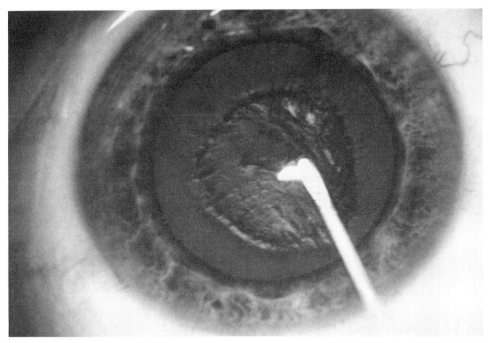

Figure 9-19. A 5.5-mm slightly eccentric but basically round anterior capsulotomy.

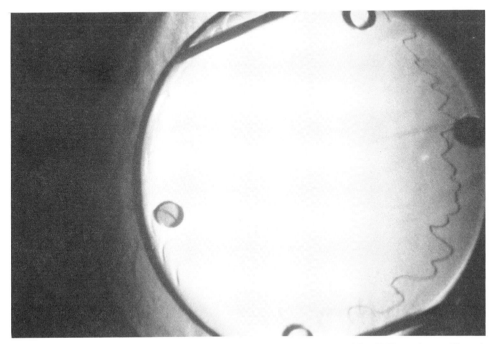

Figure 9-20. Anterior capsulotomy featured in the roundel technique of Dr. T. Hara, Japan. Note that even though the capsulotomy consists of a multiple tear pattern, no anterior radial equatorial tears exist as accomplished by Dr. Hara. (Photo courtesy of Dr. T. Hara.)

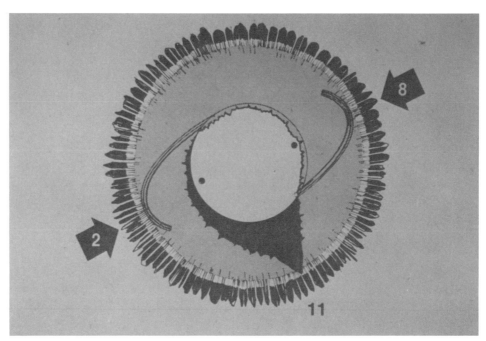

Figure 9–21. A modified J-loop IOL is placed with loops positioned 90 degrees away from the 11 o'clock tear and contained within the 2 o'clock and 8 o'clock capsular bag.

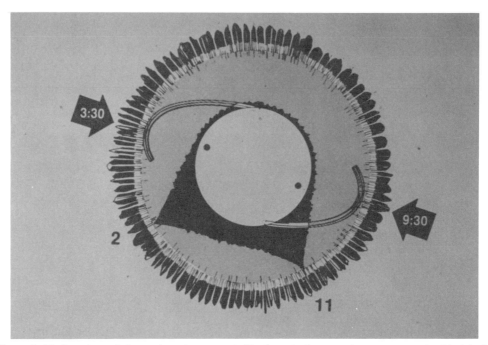

Figure 9–22. Anterior radial tears have been created in the anterior capsular remnant at the 11 o'clock and 2 o'clock position. This came from lifting the nucleus too high within a generous anterior capsular remnant. The IOL haptics have been located at the 9:30 o'clock–3:30 o'clock position to take advantage of the remaining capsular support.

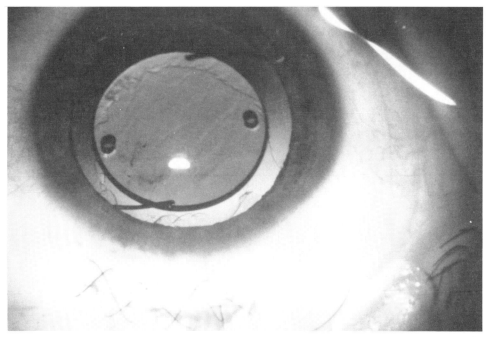

Figure 9–23. Clinical photograph of the same situation with anterior radial tears at 10 o'clock and 1 o'clock described and illustrated in Figure 9–10.

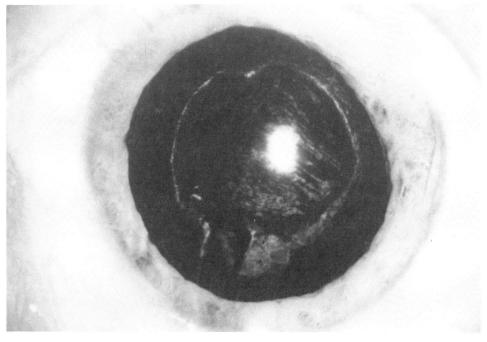

Figure 9–24. A single anterior capsular radial tear has occurred inferiorly in the roundel phacoemulsification technique. (Courtesy of Dr. T. Hara)

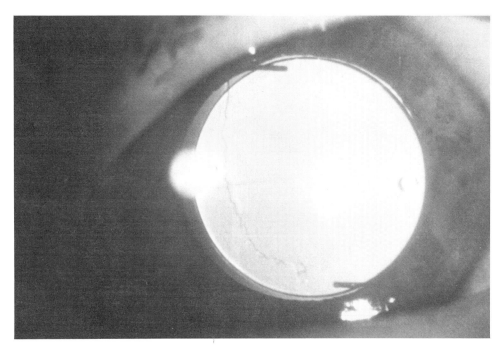

Figure 9–25. Anterior radial tear is present at the 11 o'clock position, but the IOL is well centered and contained within the capsular bag.

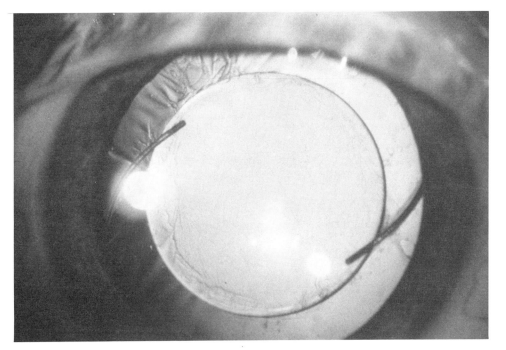

Figure 9–26. The IOL is perfectly centered within the capsular bag despite an anterior radial tear at 11 o'clock.

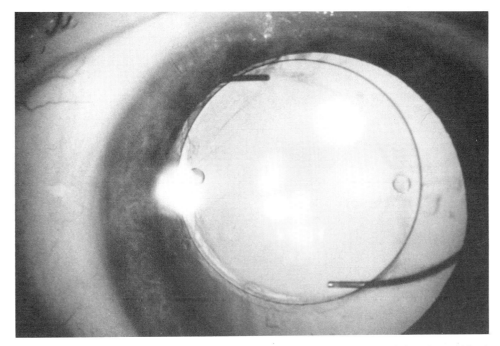

Figure 9-27. The IOL optic is decentered 1 mm to 1.5 mm superiorly with haptic location within the capsular bag. While clinically acceptable in most circumstances, subtle superior optic migration is not infrequent with a superior anterior radial tear.

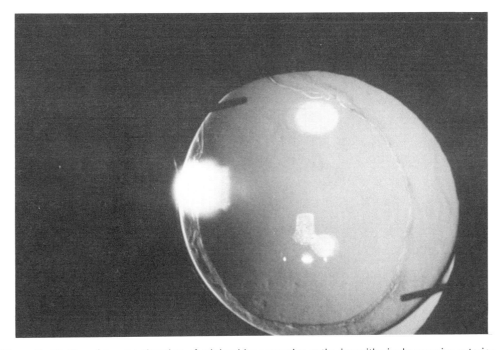

Figure 9-28. More frequent situation of minimal late superior optic rise with single superior anterior radial tear formation.

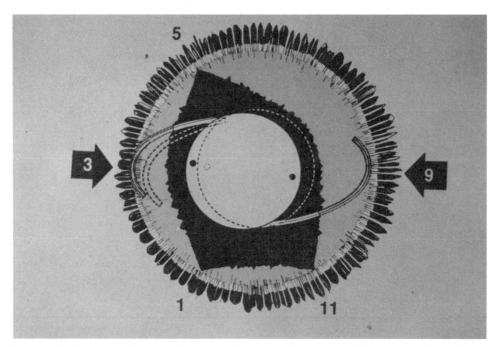

Figure 9–29. Anterior radial tear exists at the 11 o'clock, 1 o'clock, and 5 o'clock positions. There is no symmetrical support to contain a modified J-loop IOL within what is left of the capsular bag. The optic will ultimately decenter toward the 3 o'clock position because of the structural symmetrical of capsular support.

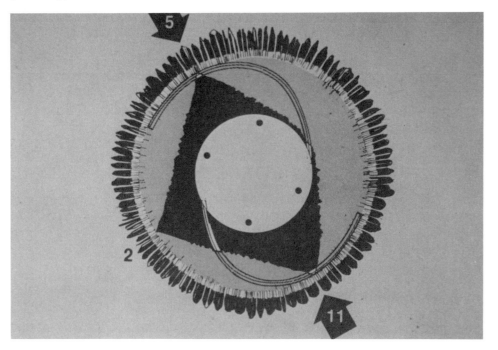

Figure 9–30. The long loops of the Simcoe–style lens bridge the anterior radial defects present in the equatorial zone and recruit to capacity for containment provided by the adjacent anterior capsular structure.

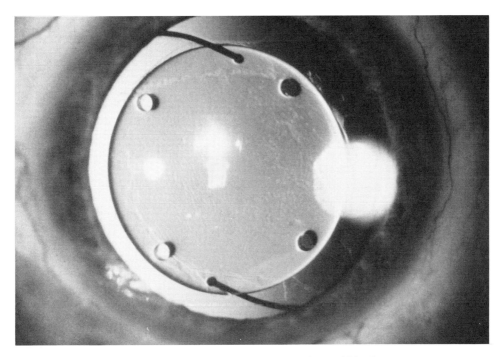

Figure 9-31. Clinical photograph of Simcoe-style lens in place within the capsular bag with two anterior radial tears present at 10 o'clock and 2 o'clock.

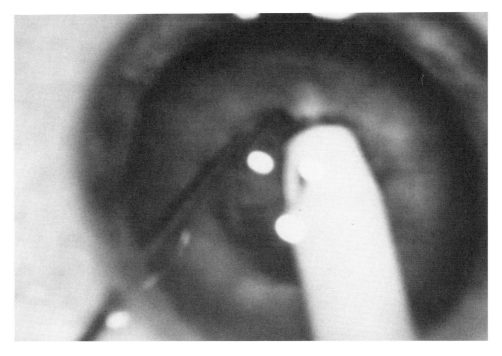

Figure 9-32. A 30-degree phacoemulsification tip burrows into the inferior nuclear rim after central sculpting is accomplished. (Photograph courtesy of Howard Gimbel, MD)

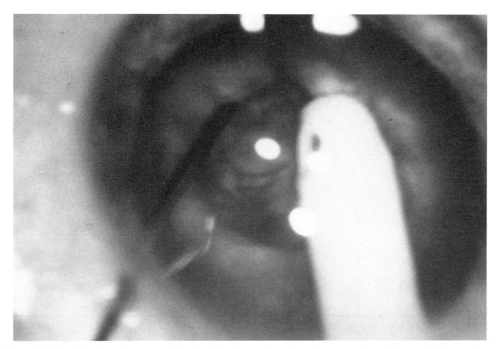

Figure 9–33. A fracture is created in the inferior nuclear rim by separating right and left sections. (Photograph courtesy of Howard Gimbel, MD)

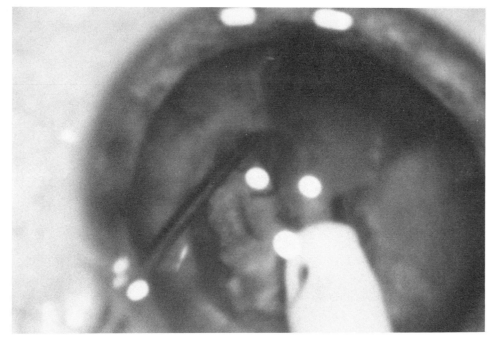

Figure 9–34. As this segment is aspirated, a second fracture develops that facilitates removal of the segment.

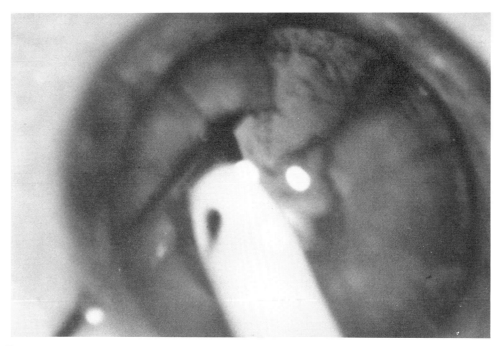

Figure 9–35. The phaco tip burrows into a segment positioned in the inferior left quadrant. (Photograph courtesy of Howard Gimbel, MD)

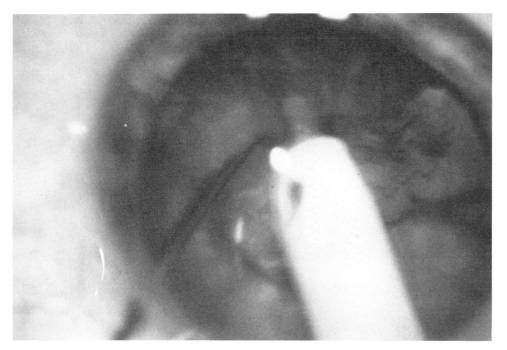

Figure 9–36. This segment is withdrawn into the tip using low power energy and suction. With rim rotation, new segments will be exposed to the inferiorly oriented tip. Rotation continues and the remaining rim and posterior nuclear disc are removed.

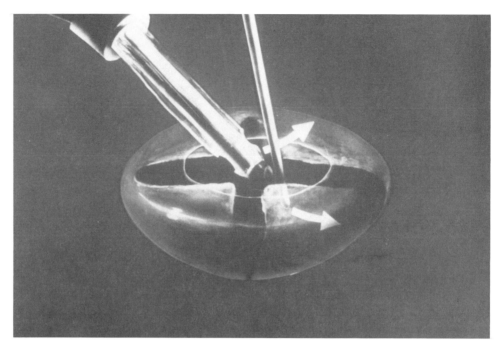

Figure 9–37. In situ phacoemulsification method of John Shepherd, MD, showing division of the peripheral nucleus into four segments and choice of attack of either inferior nuclear quadrant. (Photograph courtesy of John Shepherd, MD)

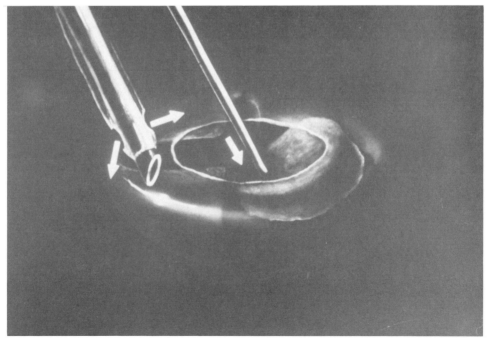

Figure 9–38. After the inferior nuclear segment is withdrawn, the two superior ones can be rotated into accessible inferior position. (Photograph courtesy of John Shepherd, MD)

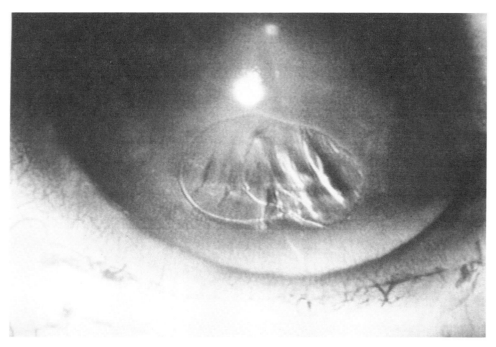

Figure 9–39. A small eccentric continuous tear anterior capsulotomy provides nuclear access during intercapsular cataract extraction, but has less of a tendency for anterior radial tear extension than a straight line or hatch mark capsulotomy. Linear capsulotomies tend to allow anterior radial tear extensions to the equator but continuous tears, even when ovoid, tend to resist such extensions. (Photograph courtesy of Marc Michelson, MD)

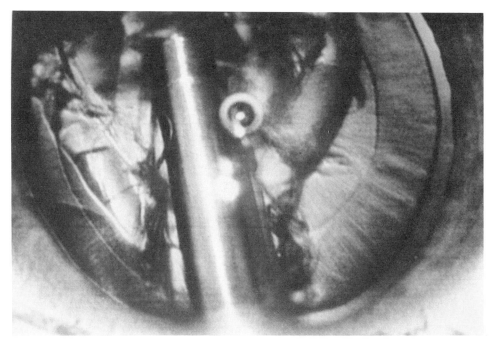

Figure 9–40. The zones of nuclear hydrodissection are particularly visible as nucleus removal progresses in intercapsular phacoemulsification. It is particularly important that the posterior nucleus be held amply above the posterior capsular surface because neither the phacoemulsification tip aperture nor its furthest posterior projection are visible. (Photograph courtesy of Marc Michelson, MD)

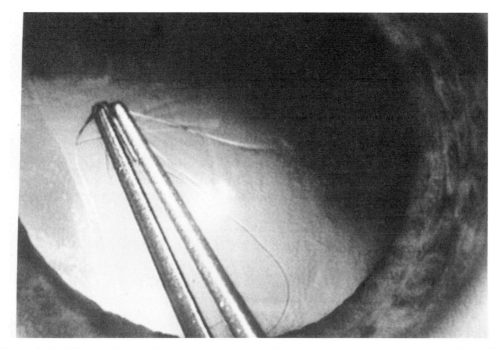

Figure 9–41. After nucleus and cortex removal, an optical anterior capsulectomy can be performed as an extension of the original mini-capsulorhexis. Although not perfectly circular, the final capsulorhexis opening will result in excellent capsular bag fixation while protecting the corneal endothelium from direct emulsification energy. (Photograph courtesy of Marc Michelson, MD)

turally perfect three-dimensional capsular envelope for placement of the intra-ocular lens. I have enjoyed these advantages, as I have been performing capsular bag phacoemulsification after capsulorhexis. The traditional capsular complications of posterior rupture or anterior radial tear are now rarely seen. I use a capsular bag technique (CBT) for more than 95% of my cases.

There are still several circumstances that I feel best lend themselves to the minimal lift superior nuclear rim prolapse technique. Some eyes with pseudoexfoliation have extremely loose zonular attachment. Because of this, it is possible to draw a loosely held flaccid capsule into the phaco tip aperture if suction is applied from the central cavity of the lens toward the equatorial capsule in a CBT. A minimal lift technique will apply suction at the superior nuclear rim from the outside of the nucleus in so that peripheral capsular aspiration is not possible. For the same reason, the minimal lift technique should be used if there exists any inferior zonular weakness, due to trauma, for example. Advanced age sometimes can combine general zonular laxity with a very firm lens. In these cases, if the cataract is 3+ brownish hard, I feel that it is simply more safe and efficient to fall back to a minimal-lift superior nuclear prolapse technique.

Phacoemulsification within the capsular bag then is essentially an ultrasonic dissection and removal of the lens nucleus in situ. The purpose of this dissection is simply to remove the cataractous nucleus while preserving a structurally perfect capsular bag for intraocular lens (IOL) placement. To accomplish this dissection, it is necessary to analyze the anatomy of the lens. It is important to conceptualize the lens as having three vertical zones: thin superficial, relatively thicker (approximately 2/3 thickness) midlevel, and thin deep zones, with all three having central and peripheral horizontal latitudes. The previously mentioned nucleus relocation techniques relied on substantial three–dimensional movement with vertical manipulation of the nucleus so that the phacoemulsification tip could remain relatively stationary in or above the iris plane and new sections of nucleus repeatedly presented to it. Nuclear material was removed by driving the phacoemulsification tip through the presented nuclear material. The superficial, midlevel, and deep peripheral portions of the lens were presented to the tip by vertical nucleus relocation and manipulation.

In order to successfully employ any of the strategies or hybrid combination of strategies of ultrasonic nuclear dissection within the capsular bag, three phacoemulsification principles must be applied in a meticulously detailed fashion.

Principle 1. Isolation of phacoemulsification tip functions into three modes: manipulation, cutting, and suction.

In the manipulation mode, the tip is simply used as an instrument to push and move the nucleus or nuclear fragments. No emulsification energy or suction are applied. Using the cyclodialysis spatula in the second hand, the lens nucleus can be rotated or the posterior fibers separated, which can result in a fracture of, or crack in, the posterior nuclear disc.

In the cutting mode, nuclear material is shaved away while not permitting the phacoemulsification tip to become occluded. Shaving of material can be allowed to occur relatively close to the capsule with the 45–degree tip. With this tip design there is less danger of drawing the capsule in with uncontrolled occlusion and aspiration. This is possible because the cutting action occurs only at the end and forward edges of the tip. On the other hand, vacuum rise and the actual suction into the substance of a large plate of nuclear material is only possible when the complete aperture is occluded. Thus, cutting and suction are isolated by design of the phacoemulsification tip and the design of the operative manipulations.

In the suction mode, the phacoemulsification tip is deliberately occluded with nuclear material. When this occlusion occurs, vacuum can actually start to build and become an important factor in the physics of the operation. Prior to occlusion, the establishment of any significant actual vacuum across the tip aperture is not possible. Not considering emulsification energy, the major operative dynamic principle in the non-occluded state is aspiration flow rate. Flow rate has significance when aspirating the liquid emulsate during the cutting process. Here, flow rate is high and vacuum is low. However, when the tip is occluded, just the opposite situation is created. Flow rate virtually drops to zero and vacuum can be allowed to build. Formed nuclear material can then be slowly drawn into the tip. In lenses of any firmness, this slow aspiration is facilitated by brief taps of extremely low phacoemulsification power, perhaps only ten to thirty percent of maximum. With soft nuclei, no emulsification power is needed. As higher vacuum levels are achieved, soft nuclear material is aspirated without the boring advance that intermittent low power phacoemulsification energy provides in firm lens situations. Peripheral suction efforts generally are not applied when the tip is close to the capsule. Peripheral material is drawn centrally not only by suction and slow aspiration, but by a central movement of the tip itself. In this way, peripheral nuclear material can be aspirated when the tip is in a safer central location.

Principle 2. Nuclear separation

The nucleus is separated into more manageable portions to be attacked by the phaco tip in its cutting and suction modes. This is inherently necessary if we are to remove the lens substance through a 5.5 mm opening in the anterior capsule. It can also be accomplished by folding over and breaking off peripheral anterior and midlevel nuclear wedges with suction after thinning of the deeper nuclear plate. Bimanual separation works well with lenses with enough firmness so that a cyclodialysis spatula will not easily sink through the remaining nucleus. Peripheral nuclear collapse and wedge aspiration works well in softer nuclei.

Principle 3. Nuclear rotation.

Inferior nuclear attack is required in all cases since successful, uniform, convenient superior access is denied by the generous intact anterior capsular remnant left after capsulorhexis. The segment of the nucleus to be acted upon by the phacoemulsification tip must generally be rotated to an inferior position below the 3 o'clock-to-9 o'clock axis. This is true no matter what tip action is desired. Cutting, suction, manipulation, bimanual segmentation (cracking), wedge fracture, and rotation maneuvers involving the phacoemulsification tip are all accomplished by inferior segment manipulation. Only a two–dimensional rotation occurs. No vertical movement of the nucleus is involved in its access.

To be a component of an overall successful cataract surgical procedure, any of these phacoemulsification methods need to be complemented by meticulous wound construction and closure.[30,31] They also all require the same good attention to technical detail in accomplishing the important step of capsulorhexis as well as cortical aspiration and IOL implantation. All cataract removal techniques have a common optimal setup of the operating room equipment and assignment of surgical procedures.

The purpose of the following chapters is to outline the important details of capsular bag phacoemulsification surgical techniques. Each of these details is of critical importance. No step can be taken without the successful completion of the previous step. There is an absolute need to be obsessed with detail and accuracy. The surgeon must appreciate everything from the subtle to the obvious. The stage must be properly set and every move perfectly executed, or the operation will be unnecessarily difficult and the risk of failure will be increased. Understanding the concepts associated with these details helps make capsular bag phacoemulsification an understandable, safe, efficient surgical procedure.

REFERENCES

1. Neuhann T: Theorie and Operationstechnik der Kapsulorhexis. Klin Monatsbl Augenheilkd 190: 542–545, 1987.

2. Gimbel HV: Capsulotomy method eases in the bag PCL. Ocular Surgery News, July 1, 1985, page 20.

3. Graether JM: Continuous tear anterior capsulotomy under Healon. Ocular Surgery News, July 1, 1986, page 30.

4. Baikoff G: Insertion of the Simcoe posterior chamber lens into the capsular bag. Am Intra-Ocular Implant Soc J 7:267–269, 1981.

5. Galand A: A simple method of implantation within the capsular bag. Am Intra-Ocular Implant Soc J 9:330–332, 1983.

6. Alfano G: Pretercapsular cataract extraction with capsular enclosed implant. Am Intra-Ocular Implant Soc J 10:203, 1984.

7. Hara T, Hara T: Subcapsular phacoemulsification and aspiration. Ganka 24: 1203–1207, 1982.

8. Hara T, Hara T: Subcapsular phacoemulsification and aspiration. Am Intra-Ocular Implant Soc J 10:333–337, 1984.

9. Hara T, Hara T: Recent advance in intracapsular phacoemulsification and complete in-the-bag intraocular lens fixation. J Cataract Refract Surg 13:279–286, 1985.

10. Gindi JJ, Wan WL, Schanzlin DJ: Endocapsular cataract surgery-1. Surgical technique. Cataract 2:6–10, 1985.

11. Girard LJ: Pars plana vs. the limbal approach to intraocular surgery. Ophthalmic Surg 12:317, 1981.

12. Wilson DL, Parel J-M, Phacoexcavation as an alternative pars plana technique for lens removal. Am J Ophthalmol 100:528–533, 1985.

13. Kelman CD: Phaco-emulsification and aspiration. Am J Ophthalmol 64:23–35, 1967.

14. Colvard DM, Kratz RP, Mazzocco TR, Davidson B: Endothelial cell loss following phacoemulsification in the pupillary plane. Am Intra-Ocular Implant Soc J 7: 334–336, 1981.

15. Davison JA: Minimal lift-multiple rotation technique for capsular bag phacoemulsification and intraocular lens fixation. J Cataract Refract Surg 14:25–34, 1988.

16. Hara T, Hara T. Roundel phacoemulsification technique for in-the-bag intraocular lens fixation. J. Cataract Refract Surg 13: 441–446, 1987.

17. Davison JA: Analysis of capsular bag defects and intraocular lens positions for consistent concentration. J Cataract Refract Surg 12: 124–129, 1986.

18. Apple DJ, Park SB, Merkley KH, Brems RN, Richards SC, Langley KE, Piest KL, Isenberg RA: Posterior chamber intraocular lenses in a series of 75 autopsy eyes Part I: Loop location. J Cataract Refract Surg 12:358–362, 1986.

19. Park SB, Brems RN, Parsons MR, Pfeffer BR, Isenberg RA, Langley KE, Apple DJ: Posterior chamber intraocular lenses in a series of 75 autopsy eyes Part II: Postimplantation loop configuration. J Cataract Refract Surg 12: 363–366, 1986.

20. Brems RN, Apple DJ, Pfeffer BR, Park SB, Piest KL, Isenberg RA: Posterior chamber intraocular lenses in a series of 75 autopsy eyes Part III: Correlation of positioning holes and optic edges with the pupillary aperture and visual axis. J Cataract Refract Surg 12: 367–371, 1986.

21. Hansen SO, Tetz MR, Solomon KD, Borup MD, Brems RN, O'Morchoe DJC, Bouhaddou O, Apple DJ: Decentration of flexible loop posterior chamber intraocular lenses in a series of 222 postmortem eyes. Ophthalmology 95:344–349, 1988.

22. Davison JA: Bimodal capsular bag phacoemulsification: A serial cutting and suction ultrasonic nuclear dissection technique. J Cataract Refract Surg 15: 272–282, 1989.

23. Shepard, J: In-situ fracture phaco method. Interview. Phaco & Foldables, November/December, 1989. page 1.

24. Hara T, Hara T: Fate of the capsular bag in endocapsular phacoemulsification and complete in-the-bag intraocular lens fixation. J Cataract Refract Surg 12: 408–412, 1986.

25. Patel J, Apple DJ, Hansen SO, Solomon KD, Tetz MR, Gwin TD, O'Morchoe DJC, Daun ME: Protective effect of the anterior lens capsule during extracapsular cataract extraction. Part II: Preliminary results of clinical study. Ophthalmology 96:598–602, 1989.

26. Patel J: One-handed technique of endocapsular phacoemulsification. Ocular Surgery News. January 1, 1987, page 24.

27. Michelson, MA: Endocapsular phacoemulsification: The storm within the calm. Ocular Surgery News, September 15, 1989, pages 60–61.

28. Wan WL, Gindi JJ, Schanzlin DJ: Endocapsular cataract surgery-II. Effects on the corneal endothelium. Cataract 2: 11–14, 1985.

29. Hara T, Hara T: Clinical results of endocapsular phacoemulsification and complete in-the-bag intraocular lens fixation. J Cataract Refract Surg 13:279–286, 1987.

30. Masket S: Deep versus appositional suturing of the scleral pocket incision for astigmatic control in cataract surgery. J Cataract Refract Surg 13:131–135, 1987.

31. Shepherd John R: Induced astigmatism in small incision cataract surgery. J Cataract Refract Surg 15: 85–88, 1989.

10

Types of Endocapsular Procedures

Paul S. Koch, MD

Types of Endocapsular Procedures

The terminology in endocapsular phacoemulsification can get a bit complicated. The same steps can be described by different words, and the same words can mean different things. Dr. Fine's hydrodissection, Dr. Dillman's hydrodemarkation, my hydrodelamination, and hydrodelineation all refer to the formation of a cleavage plane between the inner nucleus and the outer nucleus.

We are tempted to standardize the terminology at this time but refrain from doing so for a singularly important reason. There is so much energy and enthusiasm driving the development of endocapsular techniques that we feel it is more important to foster ingenuity rather than to limit one's choice of terminology. It is entirely possible that a few years from now the words we use to describe each step of the procedure and, for that matter, the procedures themselves will change dramatically.

We can, however, based on our present usage, define four categories of endocapsular phacoemulsification.

1. Endocapsular Phacoemulsification Without Nuclear Cleavage

This technique removes the nucleus while working entirely within the capsular bag, but there is no attempt to separate an inner nucleus from an outer nucleus. Examples of this type of endocapsular phacoemulsification are in situ phacoemulsification and capsular bag phacoemulsification.

2. Endocapsular Phacoemulsification With Nuclear Cleavage

This technique begins with hydrodelamination, also referred to as hydrodelineation or hydrodemarkation. This step creates a cleavage plane between the firm inner nucleus and the softer outer nucleus. Phacoemulsification begins on the inner nucleus and, after completion, continues with the removal of the outer nucleus. Examples of this type of endocapsular phacoemulsification are Chip & Flip and Spring Surgery.

3. Endocapsular Phacoemulsification Using the Cracking Techniques

This technique involves sculpting a trough deep into the nucleus and then separating the nuclear halves with two instruments until the posterior plate of the nucleus cracks. The procedure often continues twice more, cracking the two nuclear halves into four nuclear quadrants. Each of the four quadrants can be pulled away from the capsular bag and removed one after the other. Examples of this type of endocapsular phacoemulsification are Fractional 2/4 Phaco, Divide and Conquer, and Maltese Cross.

4. Intercapsular Phacoemulsification

This technique involves keeping most of the anterior capsule intact, making only a small capsulorhexis superiorly and working through it. It is not necessary to make the smallest capsulotomy possible, only a small one that leaves most of the anterior capsule intact. That way, the nucleus can be emulsified within a nearly intact capsular bag, isolating the action from the anterior chamber and the cornea.

Section V

ENDOCAPSULAR PHACOEMULSIFICATION WITHOUT NUCLEAR CLEAVAGE

11

In-Situ Phacoemulsification

William J. Fishkind, MD

In-situ phacoemulsification is a technique made possible by the explosion of technological improvements in phacoemulsification. Because the surgeon could not control power in the early phaco machines, it was necessary to prolapse the lens nucleus into the anterior chamber, or at least into the pupillary plane, to protect the delicate structures of eye from the explosive power of the poorly controlled phaco tip. New generation machines with exquisite linear control of power and greater flexibility have allowed surgeons to perform techniques of phacoemulsification that were impossible prior to this time.

Among these new techniques is in situ phacoemulsification. In this technique, the majority of phacoemulsification is performed deep to the pupillary plane using low, linear power. The technique is one-handed and involves a slow and deliberate phacoemulsification. This procedure does require the latest generation machine and linear power. By using this technique, phaco power is placed far away from the corneal endothelium, thus preserving this delicate structure. In addition, at the low powers used for this procedure, damage to the posterior capsule and equatorial capsule is rare.

This procedure is indicated for all types of cataracts except very hard ones. It can be used in patients who are old and in those with pseudoexfoliation, previous trauma, and weak zonules. It is contraindicated in the hard nucleus; the peripheral ring of lens tissue is too hard to collapse, a critical part of this procedure. In these cases, a two-handed, standard phacoemulsification at the superior pole, or a Divide and Conquer technique is recommended.

In situ phacoemulsification begins with a keratome incision followed by a 5 mm capsulorhexis (Figures 11–1, 11–2). The machine is set at 60% to 70% power for most nuclei. If the nucleus is very soft, lower power, i.e., 40% to 50%, is recommended. A 30– or 45–degree phaco tip can be used, but this procedure is best performed with a 30–degree tip because the aspiration qualities of the tip are better and good aspiration is essential to the successful outcome of this procedure.

Hydrodissection may be carried out if the nucleus is soft. It is probably not necessary if the nucleus is moderate to hard, but it can be performed if the surgeon is more comfortable with this technique and wants to ensure that the nucleus will separate from the posterior capsule. If the nucleus is very hard, hydrodissection is contraindicated because it will allow the nucleus to become free too quickly, causing loss of control of the nucleus during phacoemulsification.

Sculpting is begun superiorly using low power on the surface of the lens. This material is transitional nuclear cortical material and does not demand high power (Figure 11–3). As the phacoemulsification is carried toward the center of the lens, the fetal nucleus is encountered. This nucleus is much harder than the surrounding tissue and, therefore, needs increased power (Figure 11–4). As the surgeon passes through the fetal nucleus, the material begins again to change to transitional, nuclear-cortical material, so power should be gradually decreased as one approaches the plane of the posterior capsule (Figure 11–5).

The sculpting is carried as widely and as deeply as prudently possible, moving from twelve o'clock to six o'clock in an orderly fashion (Figure 11–6). As noted above, power should be decreased near the posterior capsule and in the equatorial regions. The phaco tip may pass through the peripheral cortical material but will not break the posterior capsule if power is low.

When maximal sculpting has been performed in the twelve o'clock to six o'clock axis, the phaco tip is used to engage the nuclear rim at nine o'clock. Gently pushing in an arcuate fashion, the nucleus is rotated 90 degrees to the left. This moves the nine o'clock nucleus to the six o'clock position (Figure 11–7). Once this is done, there is new material now available at six o'clock, and sculpting is continued moving from twelve o'clock to six o'clock, again, in an orderly fashion, creating a very thin, nuclear bowl (Figure 11–8).

When no more material is present at six o'clock, the lens nucleus is once again rotated 90 degrees to the left, bringing what was originally at twelve o'clock, now at nine o'clock, to the six o'clock position (Figure 11–9). Sculpting is continued from twelve o'clock to six o'clock, removing all new material at the six o'clock position (Figure 11–10).

At this point, the nucleus should be reduced to a thin shell. It should be possible, in position two, using aspiration only, to grasp the thin rim of nuclear material that lies just under the anterior capsular remnant to six o'clock (Figure 11–11). Once grasped, it is pulled gently toward the center of the pupil, and low power phacoemulsification is performed, thus, in essence, "splitting" the nucleus from its attached nuclear rim (Figure 11–12).

The nucleus can then be rotated left or right, and the next piece of nuclear tissue is pulled to the center of the pupil at the pupillary plane for phacoemulsification and removal (Figure 11–13). At this point, there is only a small amount of nuclear material remaining, which will, we hope, follow and be easily removed with minimal phacoemulsification.

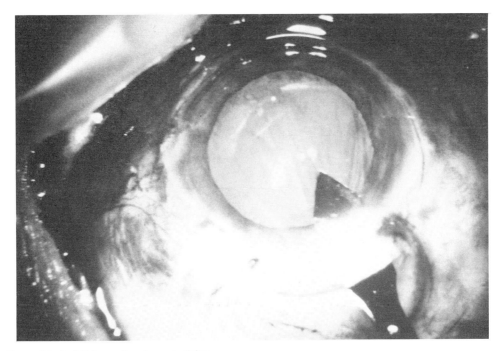

Figure 11–1. A 3.2-mm keratome incision.

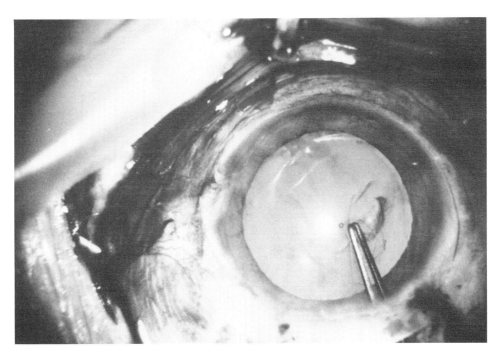

Figure 11–2. A 5-mm capsulorhexis.

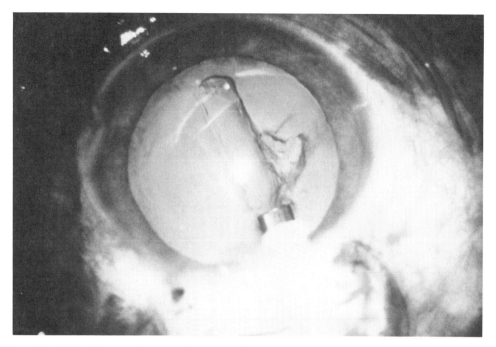

Figure 11-3. Sculpting begins using low power because the first layer of tissue is soft, transitional material.

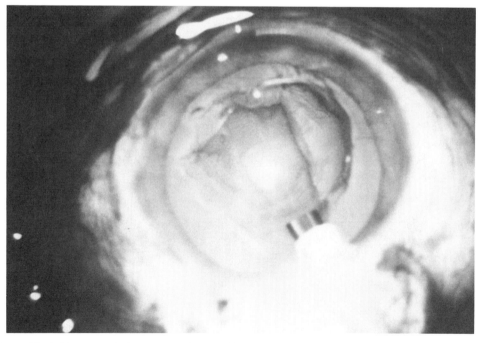

Figure 11-4. Deeper sculpting passes through the fetal nucleus, which is much more dense. More ultrasound power is needed here.

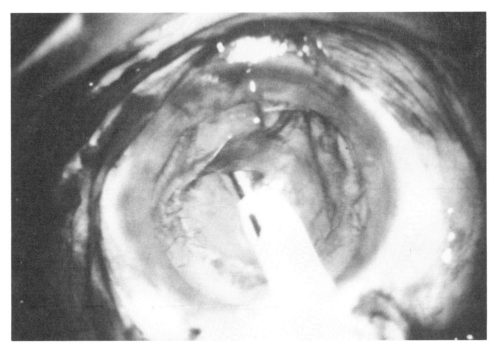

Figure 11–5. Very deep sculpting through the fetal nucleus and back into the soft transitional material. Less ultrasound power is needed. The red reflex becomes much brighter.

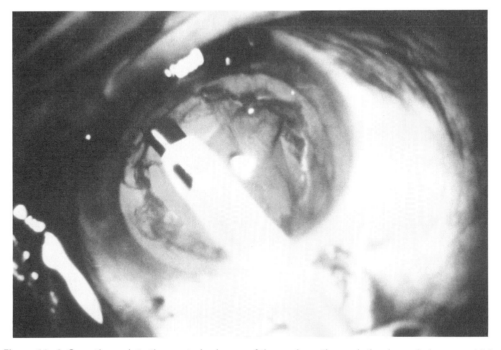

Figure 11–6. Once through to the posterior layers of the nucleus, the sculpting is carried out as widely and as deeply as possible.

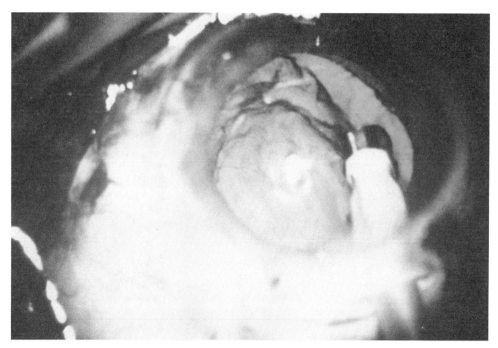

Figure 11–7. After maximum sculpting, the phaco tip engages the nucleus at 9:00 o'clock and rotates that point to the 6:00 o'clock position.

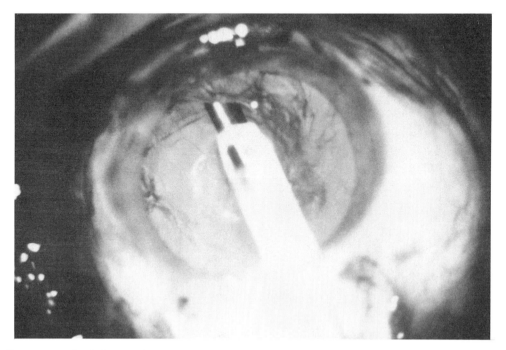

Figure 11–8. Sculpting continues at 6:00 o'clock, removing the tissue that had been rotated in from the 9:00 o'clock position.

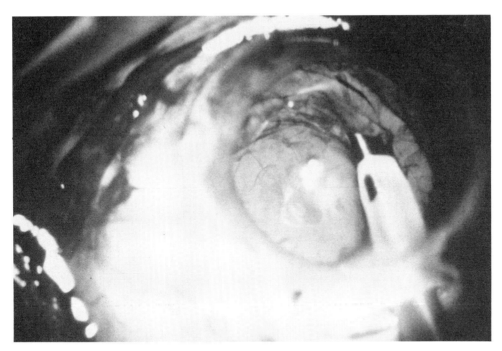

Figure 11–9. Further rotation of tissue from the 9:00 o'clock position to the 6:00 o'clock position.

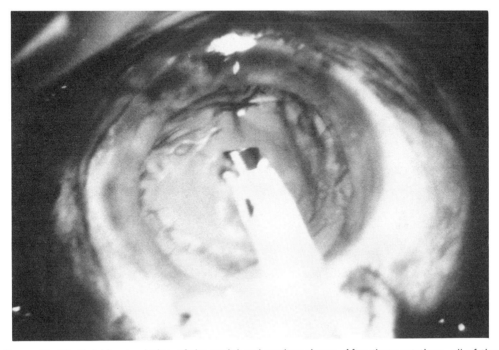

Figure 11–10. Finishing sculpting of the peripheral nuclear tissue. After three rotations, all of the nucleus has been scuplted at the 6:00 o'clock position.

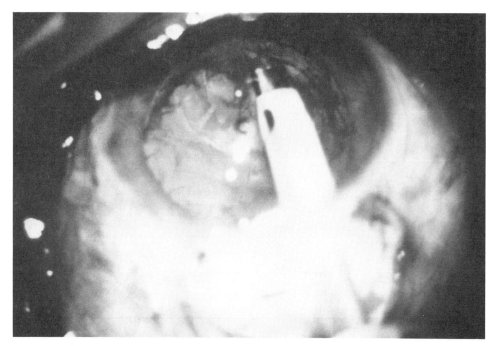

Figure 11–11. Engage the thin nuclear shell at 6:00 o'clock with aspiration, and pull it away from the capsule.

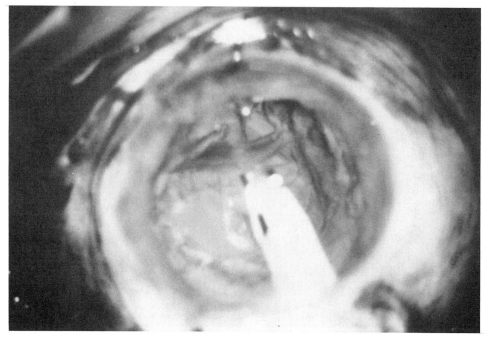

Figure 11–12. Engaging the pulling a thin nuclear rim will permit it to "split" away from the capsule.

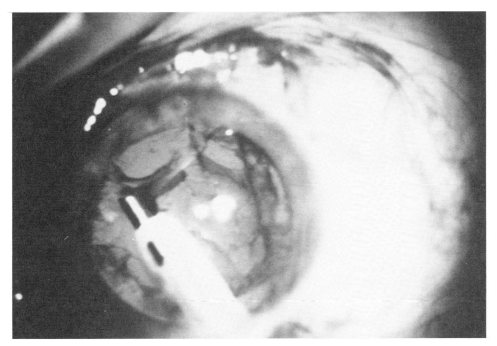

Figure 11–13. After splitting the nucleus from the capsule, it can be pulled away and emulsified in a safe zone away from the capsular bag.

It may be helpful at this point, when the nuclear rim and plate is minimal, to switch to pulse phaco. This maintains a deep anterior chamber, allowing greater surgeon control over the posterior capsule.

If it is impossible to grasp the nucleus at six o'clock under the anterior capsular rim, the center point of the nucleus can be engaged over the posterior capsule and pushed toward six o'clock. This will push the nuclear bowl ahead of it, and the nucleus will fold upon itself and present at six o'clock for easy phacoemulsification.

Problems can occur if the nucleus is very soft and will not dislocate or rotate so that the nine o'clock nucleus can be moved to the six o'clock region. In this situation, a cannula can be used to hydrodissect under the anterior capsular rim for 360 degrees. It may be necessary to move the cannula to different clock hours to provide a complete hydrodissection. This will allow the nucleus to easily rotate so that the surgeon can complete the procedure. If rotation still appears impossible, the nucleus must be extremely soft. In this situation, conversion to an 0.3 or 0.5 mm irrigation and aspiration tip should allow successful removal of the remaining nucleus.

If the nucleus is very hard and will not collapse, the surgeon should proceed by rotating the nine o'clock to the six o'clock position, as previously discussed, and continuing sculpting, little by little, more widely and more deeply, until the nucleus is thin and the red reflex is very bright. At this point, the nuclear rim should collapse and allow further removal in an in situ technique. If this is impossible, the surgeon would be wise to convert to two hands and prolapse the superior pole removing it in a standard, two-handed fashion.

This procedure should be considered an advanced technique but one that with some diligence and practice is easily within the reach of the average phaco surgeon. The technique does demand linear phaco, and it should be performed in a slow and controlled fashion. The results are extremely satisfactory with quiet eye and clear corneas due to low phacoemulsification power, decreased turbulence, and absence of any steps taking place near the corneal endothelium.

12

Capsular Bag Phacoemulsification

James A. Davison, MD

Strategy

I have selected a specific phacoemulsification strategy by this point. It is rare to require a nucleus expression or intracapsular cataract extraction, but these situations still come up and it is good to know how to perform these procedures well. Flexibility is the key to applying the correct operation to every circumstance. Each patient is different, and the surgeon should be able to match each one with a specific operative plan that, by making the operation safe and efficient, will maximize the chance for a successful outcome. Patient cooperation, pupil size, zonular integrity, and nuclear hardness all influence my decision in selecting an operative plan. In phacoemulsification, I first must decide whether to use one of three basic capsular bag techniques or a minimal–lift technique.

Minimal–Lift Technique

If the patient tends to be less than perfectly cooperative, I may lean toward a minimal–lift technique since it is faster and generally not as exacting as the capsular bag methods. This, by the way, is probably one of the best methods to begin with when starting phacoemulsification. I learned the method initially from Mike Colvard, Tom Mazzocco, and Dick Kratz. It is the method that Bill Maloney has taught to so many so well in his phacoemulsification seminars throughout the world.

If the pupil is 4 mm or smaller, I may elect to perform a sector iridectomy (depending on lens hardness) and do a can–opener capsulotomy with a combination of minimal–lift and capsular bag methods, again depending on lens hardness. If the inferior zonule has a focal defect, I will try to perform a minimal–lift method that places most of the stress of phacoemulsification on the zonule in the superior quarters. If the superior zonules are loose, I may use a capsular bag method where the intensity of effort is directed inferiorly. If diffuse weakness exists, as in some cases of pseudoexfoliation or extreme old age, I may use a minimal–lift technique, which directs dissection and aspiration from the outside in rather than from the interior of the lens out to the capsule and weakened zonules, where I might encounter capsular aspiration.

The only disadvantage to the minimal–lift method is the single anterior radial capsular tear, which is almost always produced at the 11 o'clock position. This is still an excellent and very safe technique and should be used in difficult circumstances or if one is just learning phacoemulsification (Figures 12–1 through 12–5).

Capsular Bag Phacoemulsfication

At this point in my surgical evolution, I use three strategic variations of capsular bag technique, depending on lens firmness. There exists a fair amount of overlap so that, in most cases, either of two adjacent or many times all three techniques could be appropriately selected. Additionally, various features of one strategy can at any time be integrated into another; there is a great deal of flexibility to handle virtually any clinical circumstance.

1. Soft lenses–I usually select my basic cutting, suction peripheral nuclear collapse technique as my preferred method here.

2. Medium firm lenses–A multiple wedge separation technique can be used very easily here. This resembles a modification of Howard Gimbel's Divide and Conquer technique except that the separation is done with cutting and aspiration rather than the cyclodialysis spatula. Adjacent methods one or three also can be used for these lenses.

3. Firm lenses–A posterior nuclear plate separation quartering process is used here to facilitate infolding and aspiration of the nuclear periphery. The posterior nuclear cracking maneuver has been adapted from that described by Howard Gimbel and John Shepherd. The separation of the posterior nucleus into quartered segments has been integrated into the suction and cutting theme used for softer lenses so that hard lenses can be easily and efficiently emulsified just as soft lenses are without it.

I call this the "CCS hybrid technique" (Cracking, Cutting & Suction) because it incorporates the principles of independent suction and cutting nuclear dissection along with the valuable facilitative maneuver of posterior nuclear cracking needed for easy emulsification of firm lenses.

Basic Preliminaries

Certain preliminaries are common to phacoemulsification in general.

Eyes with 2+ nuclear yellowing and firmness and a surrounding softer nuclear and cortical layers will be the easiest to do by any of the outlined methods (Figure 12–6).

The machine must be tested and ready. Back-up handpieces, footpedals, packs, tips, and tubing are essential. An entire back-up machine is ideal. You must know the complex control panel functions and mechanics of the machine intimately so that you can direct trouble shooting efforts and solve problems (Figure 12–7). You must be able to tell your circulating nurse how to solve technical problems even though he or she has had in-service training.

When you are ready, receive the handpiece from your scrub nurse. The test chamber cap should be place so that you can run test the handpiece in your own hands in one last–minute check (Figure 12–8). Make sure that both the irrigation and aspiration lines are firmly attached. A catastrophe may occur if the irrigation line falls off during phacoemulsification. Listen to the ultrasonic energy. At 100% energy, it should sound like a "normal" 100%, not weak or variable. If it does not sound right, check the tightness of the tip and retune. If this is not effective in achieving perfect sound, change tips or handpieces. Also check to make sure the footpedal is responsive to "surgeon control." Look at the silicone sleeve. Is the correct amount of tip showing? Are there any defects in it?

Make sure that rectus sutures are released so that you can move the eye freely. Your mask is tight so that your oculars don't fog. Make sure that you are comfortable. Your lower legs should be extended slightly so that you can make the delicate movements needed with your feet without loosing balance or moving your trunk.

Grasp the anterior sclera at the wound and lift it slightly as the tip is inserted into the anterior chamber. The tip should be bevel down with and the infusion should be on (foot position 1) (Figure 12–9). This approach creates the least trauma to Descemet's membrane and the iris. If the iris is traumatized here, it will have greater tendency to prolapse later. Turn the tip over and briefly tap into foot position 2 to aspirate the remaining viscoelastic and the anterior capsule remnant.

Sculpting is begun in shallow fashion using fairly low power. Unnecessary high power at this point creates sudden turbulence underneath the iris. This ultrasonic trauma causes immediate loss of significant pupil size and makes the rest of the operation difficult. Nuclear material is shaved away without permitting the tip to become occluded. Varying amounts and patterns are removed depending on nuclear hardness and the ultimate strategy to be used.

Stability of the phacoemulsification tip and handpiece control are created through comfortable hand positioning that will provide good support and positional versatility (Figures 12–10, 12–11). The index fingers of the right and left hand work together to guide the tip. The handpiece is supported mostly by the right fingers and is stabilized by the left. Both hands, fingers, and arms work together as a single unit. At times, the index finger of the left hand is on the left side of the tip; at other times, it is on its top. The three lines leaving the handpiece are draped over the right forearm so that no traction or gravity acts on them (Figure 12–12).

As sculpting descends further into the midnuclear level, an irritating reflex may appear as the microscope light reflects back from the surface of a lake of pooled BSS (Figure 12–13). This occurs especially in eyes with deeper orbits. The reflex is bad enough, but if the fluid rises enough so that a portion of the peripheral cornea is covered, a dangerous loss of depth perception occurs. Ribbons of precut gelfoam may be placed in one or both fornices to wick away excess fluid (Figure 12–14).

Changing phaco tip orientation and fine manipulation of nuclear material are characteristics of any capsular bag technique. As the right side of the interior nuclear bowl is shaved, the handpiece is rotated bevel open to the left so that the opening faces the nuclear center (Figure 12–15). This tip rotation is generated by a change in finger and wrist position (Figure 12–16). When the tip then shaves the left nuclear interior surface, the wrist and finger position change to rotate the tip so that it is open with its bevel to the right. The wrist and forearm can only suppinate so much before the action becomes awkward and tip control may become less than optimal. A repositioning of the hands and fingers to the original position with the tip in its new orientation produces much finer control.

When performing dissection of deeper or more superior nucleus, the tip is aimed more posterior as the left-hand index finger acts as an important fulcrum (along with the brow on the underside) to stabilize this movement (Figures 12–17, 12–18).

It is important to use a 45–degree angled tip for capsular bag nuclear dissection. As the tip is angled deeper into the nuclear bowl, its functional angles of attack steepens. That is, when phacoemulsification begins, the tip's angle of attack is flat and parallel to the horizontal plane of the lens. In this early stage, it is easy to see the entire aperture of the tip and appreciate its shaving effect on the nucleus. As the tip probes more vertically, aperture appreciation is reduced because of the apparent overhang of the anterior aspect of the tip, which obscures the view of the working posterior edge (Figure 12–19). This critical image of the cutting edge of the operation is lost earlier with the 30–degree tip and even earlier with the 15–degree tip.

The visibility and separation of emulsification energy and suction afforded by the 45–degree tip is especially important when cutting the posterior inferior grooves in CCS hybrid phacoemulsification. The adjacent nuclear tissue on the right and left side of the grooves, which has not been cut, affords little clue as to the depth of the ongoing cut of the tip within the central grooves. Not only is it easy to cut too deep, but full suction might be applied inadvertently to the posterior capsule if 15– or 30–degree tips are used (Figures 12–20, 12–21).

When both hands are employed, the right guides the phacoemulsification tip and the left manipulates the cyclodialysis spatula (Figures 12–22, 12–23). The small and ring fingers of either hand make contact with the patient's forehead and is stabilized by the contact.

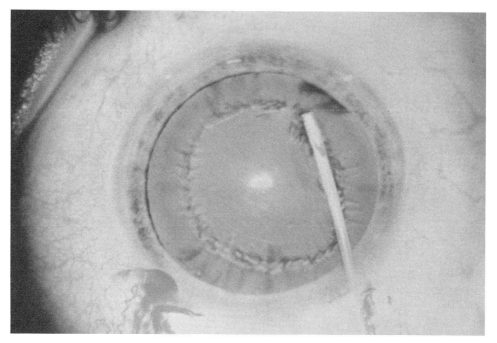

Figure 12–1. For minimal–lift phacoemulsification, a can-opener capsulotomy is created with the opening slightly eccentric toward the 11 o'clock position. This eccentricity facilitates formation of the anterior radial tear that is usually necessary to perform the MLT here.

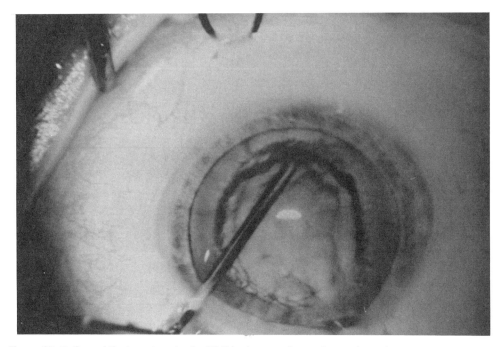

Figure 12-2. One of the key steps in the MLT is the superior nuclear pole prolapse. The pole is every so slightly brought up and only the anterior portion of the superior rim emulsified. The inferior pole is being pushed inferior and posterior by the spatula so that the superior pole can be pushed forward by relatively increased vitreous pressure. This increase is created by stopping the infusion in foot position 0 temporarily. The phaco tip is withdrawn just enough so the superior nuclear pole can come forward slightly, held in place by the spatula and emulsified by the tip.

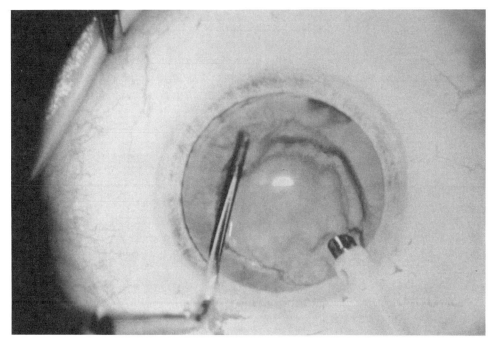

Figure 12–3. The superior anterior rim has been reduced and the nucleus has fallen back. It is then dislocated with the cyclodialysis spatula and rotated clockwise several clock hours.

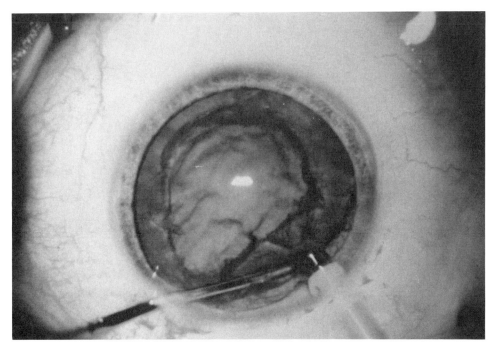

Figure 12–4. New superior rim tissue is removed after prolapse and spatula stabilization have been accomplished.

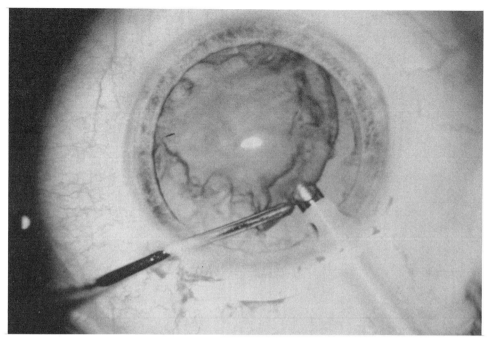

Figure 12–5. As the peripheral anterior nuclear volume is reduced, more anterior and midlevel nuclear rim can be removed with greater safety because it can be positioned in a safer more central position.

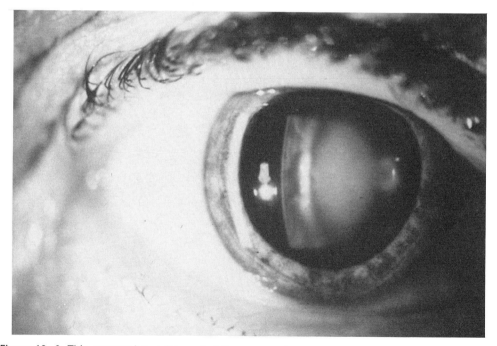

Figure 12–6. This cataract has a 2+ central firmness and a generous surrounding pillow of softer nucleus and cortex. Any phacoemulsification method will work well in these eyes. This type of cataract is ideal for the surgeon who is just beginning phacoemulsification. It should be the type sought for technique expansion and improvement by the more experienced surgeon as well.

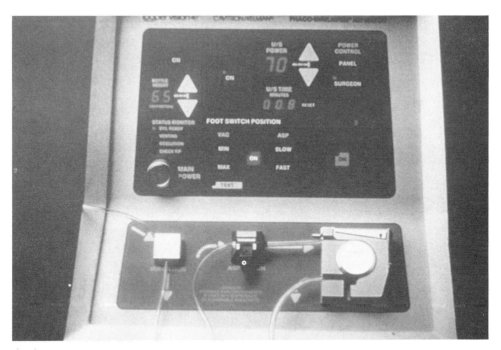

Figure 12–7. Front panel of the Alcon/Coopervision 9001 phacoemulsification machine. The surgeon and circulating nurse need to have the machine's behavior and displays memorized so that its operation can be insured and problems can be resolved quickly.

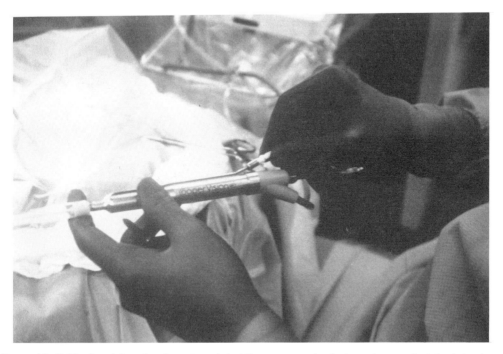

Figure 12–8. The handpiece has been tuned, but the surgeon checks to make sure that the irrigation and aspiration tubing is firmly attached while listening to the phacoemulsification tip vibrate to make sure it sounds appropriate. Poor sound should mean retightening the tip and retuning it before surgery is begun. Marginal tip performance does not improve as the operation proceeds; it usually gets worse. Loss of the irrigation line during surgery could be catastrophic.

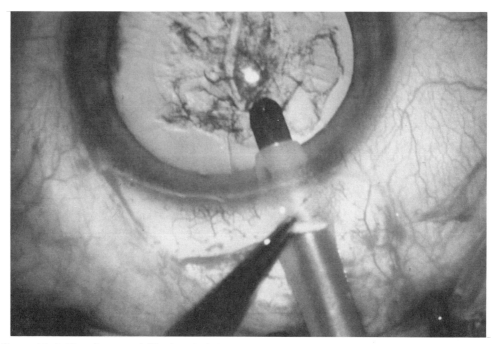

Figure 12–9. The phacoemulsification tip is inserted upside down and turned bevel up while in foot position 1, being very careful not to disturb Descemet's membrane.

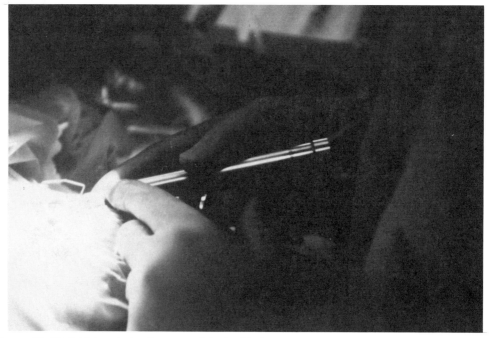

Figure 12–10. Initial sculpting is accomplished with the handpiece almost parallel to the iris. Both index fingers guide the phaco tip handpiece with the surgeon's right index finger pointing toward the action. The right hand is slightly supinated and the left slightly pronated with the index fingers in a very slightly similar position.

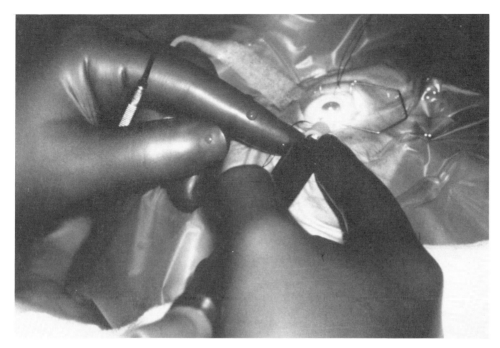

Figure 12–11. Sometimes the index fingers are just on the right and left side of the handpiece.

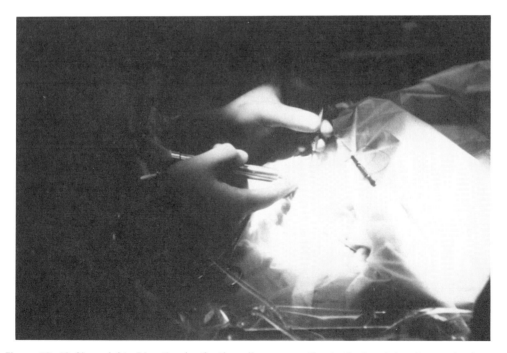

Figure 12–12. No weight of traction by the three lines connecting to the handpiece is permitted. The surgeon's hands are resting comfortably on the patient's forehead. The phacoemulsification tray is at the appropriate height with the lines in the plastic bag. The hemostat is clipped to the drape but the rectus suture has been detached.

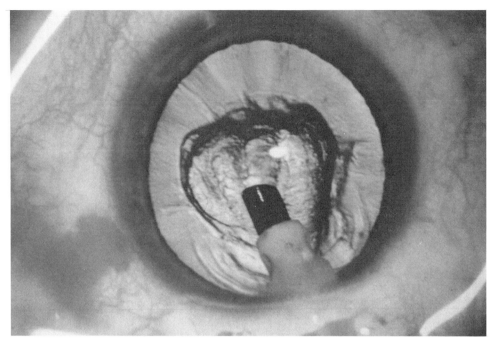

Figure 12–13. Initial sculpting is almost completed in a fairly soft lens. If too much more sculpting is done, the periphery will be too thin to allow effective manipulation. A troublesome fluid pool has accumulated and caused an irritating reflex.

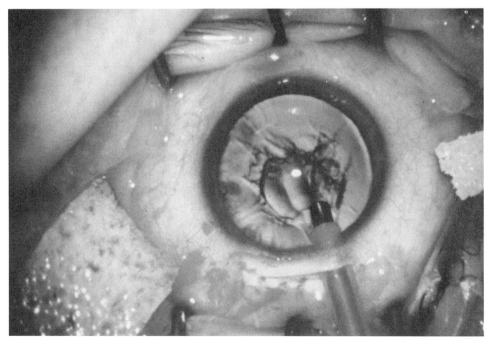

Figure 12–14. This fluid accumulation reflex problem is solved by placing one gelfoam ribbon drain in each canthal corner to remove the fluid and maintain good visualization.

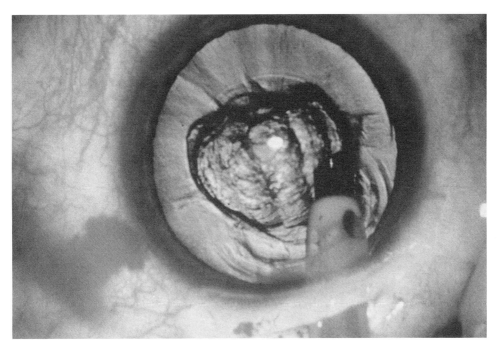

Figure 12–15. Side sculpting is accomplished with a gentle rotation of the tip so that the bevel always stays toward the central cavity created by nuclear previous nuclear sculpting.

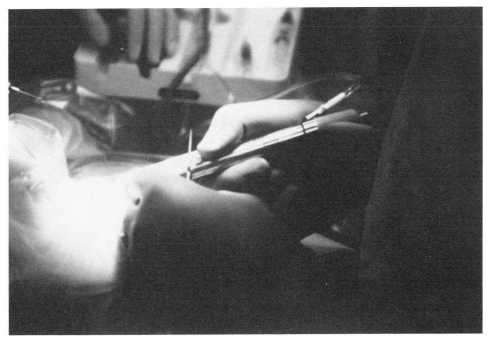

Figure 12–16. The handpiece has been rotated so that the bevel is open to the left, and the right side of the nuclear wall can be shaved. Compare the orientation of the phacoemulsification cord to the initiating position in Figure 12–10.

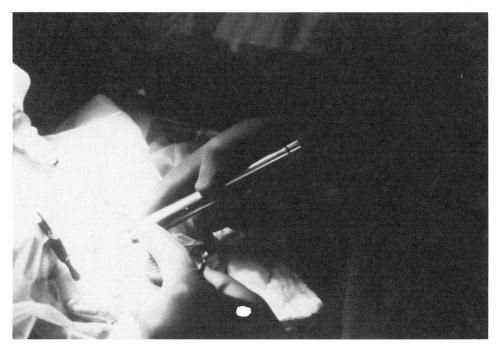

Figure 12–17. As deeper phacoemulsification occurs, the surgeon's right hand actually lifts away from the patient's forehead, but is steadied by the left index finger.

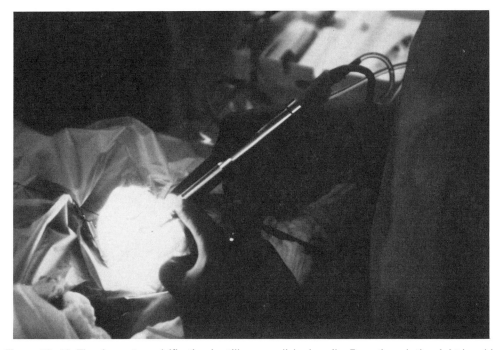

Figure 12–18. The deepest emulsification is still accomplished easily. Even though the right hand is off the forehead, it is stabilized by the left hand through the index fingers at the tip.

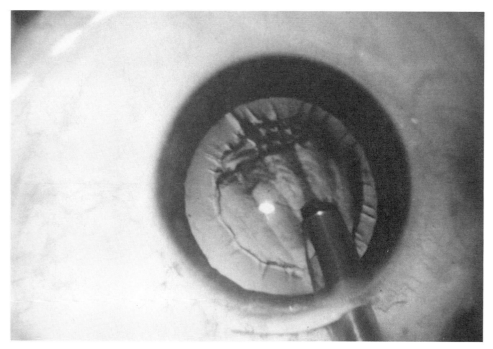

Figure 12-19. A 30-degree phacoemulsification tip allows for easy occlusion but poor visualization of deeper shaving maneuvers.

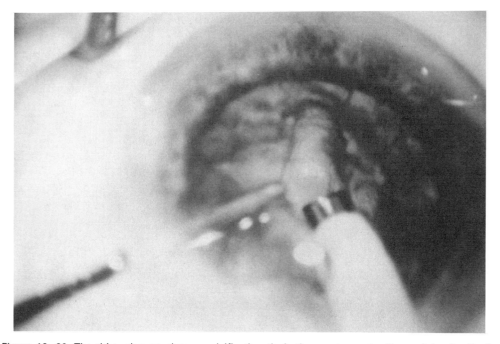

Figure 12-20. The thirty degree phacoemulsification tip is the most popular tip used, but it affords little view of the aperture or posterior edge during deeper phacoemulsification. (Photograph courtesy of David Dillman, MD.)

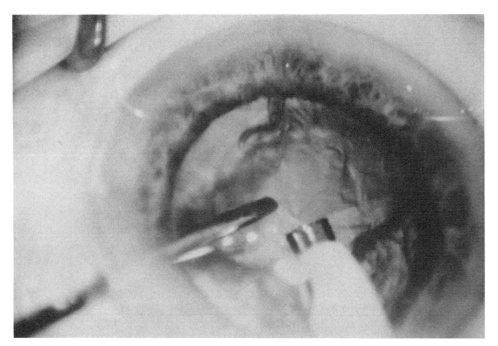

Figure 12–21. As emulsification proceeds, the 15 or 30 degree tip has a greater chance of inadvertently aspirating the posterior capsule because its tip aperture is almost facing the capsule's concave surface. (Photograph courtesy of David Dillman, MD.)

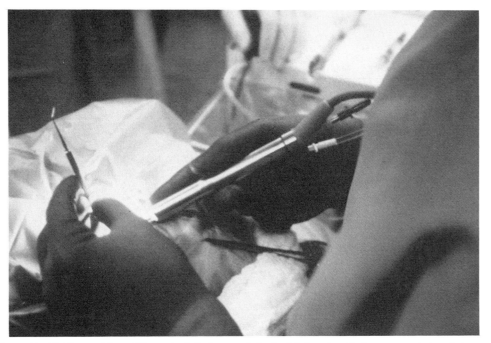

Figure 12–22. The position of the cyclodialysis spatula in the left hand, with the surgeon's hand almost completely off the forehead. The right hand directs the phacoemulsification handpiece while only the index finger of the right hand touches the brow.

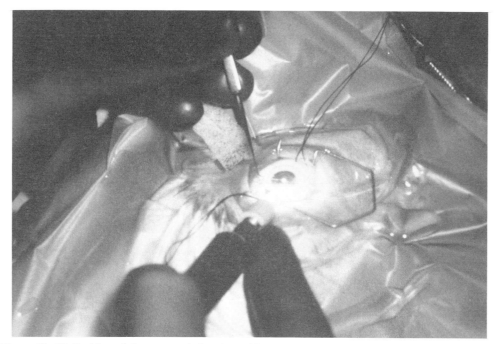

Figure 12–23. Position of the surgeon's hands and fingers as seen over his right shoulder.

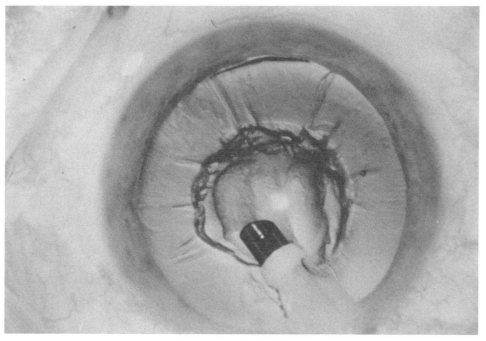

Figure 12–24. Shallow sculpting under low power is performed initially with soft nuclei.

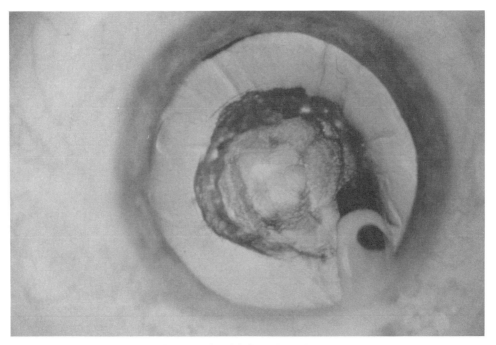

Figure 12-25. The tip is turned slightly on its side in order to keep the bevel open toward the center of the lens cavity. This helps prevent inadvertent occlusion of the tip by soft nuclear material.

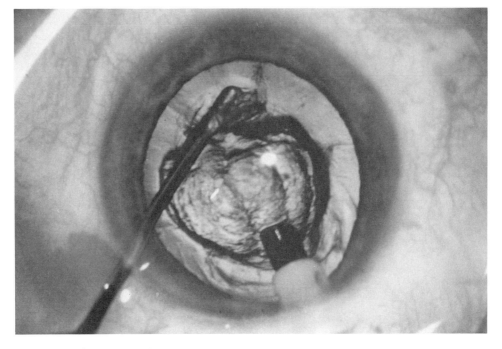

Figure 12-26. If hydrodissection has been incomplete, a nucleus dislocation can be completed with a cyclodialysis spatula by applying posterior pressure to the nuclear rim.

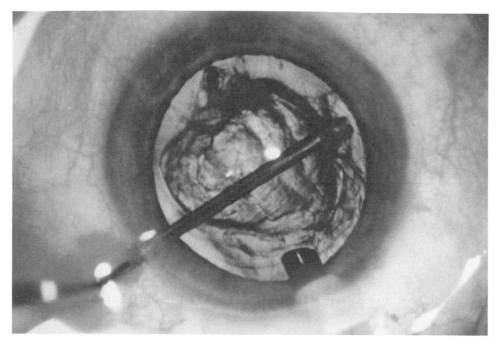

Figure 12–27. With soft lenses, great care must be taken not to sink through the nucleus in through the posterior capsule.

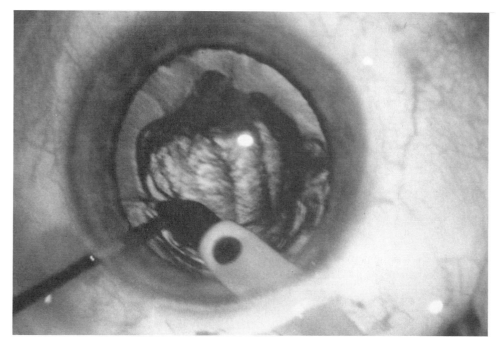

Figure 12–28. Dislocation of the superior nucleus can be accomplished with gentle inferior posterior pressure from the phaco tip. (Foot position 1)

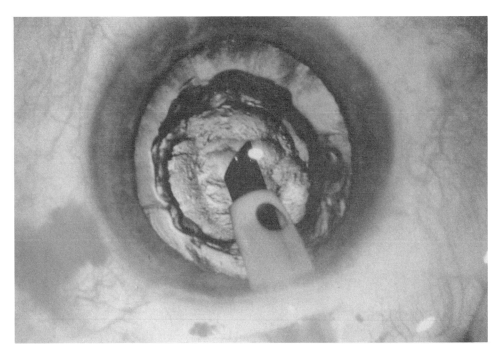

Figure 12–29. Deeper mid–peripheral nuclear thinning is being accomplished now that the nucleus has been loosened completely with the cyclodialysis spatula. The tip is on its side so that the surgeon can see the exact depth of the cut as he makes it.

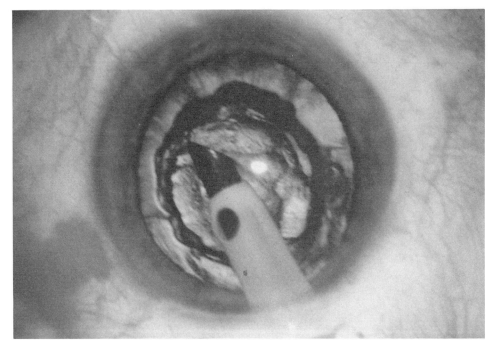

Figure 12–30. The tip is rotated so that its aperture is opened to the right as well.

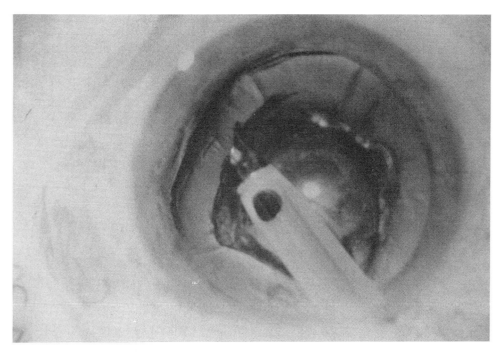

Figure 12–31. The maximum possible phacoemulsification energy has been reduced to 60% with the machine maintained on the linear control mode. Brief taps of very low power of phaco energy will start to fold in one section of soft nuclear rim.

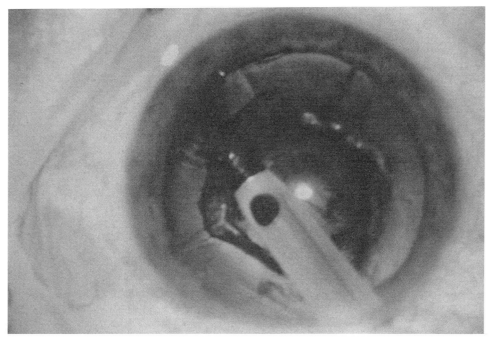

Figure 12–32. An initial plug of peripheral nuclear rim has been withdrawn by the suction from the phacoemulsification tip.

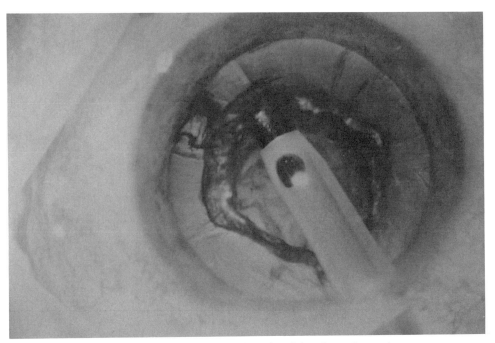

Figure 12–33. The tip is poised to cut another wedge of peripheral anterior nucleus away.

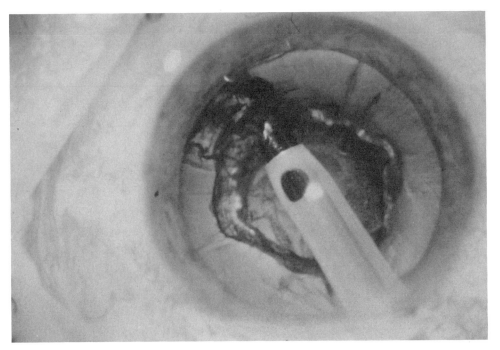

Figure 12–34. The adjacent second plug is withdrawn. This is accomplished by complete tip occlusion and tiny bursts of extremely low emulsification power. Deeper more peripheral nucleus will not follow inadvertently, because the aperture is open to the center.

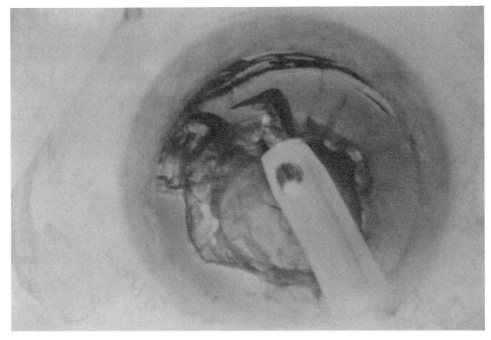

Figure 12-35. A third wedge of peripheral nucleus is removed.

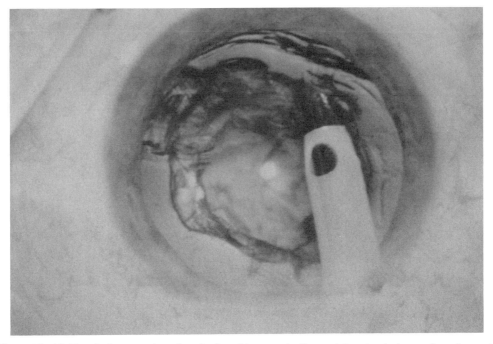

Figure 12-36. The tip is rotated so that the bevel is open to the peripheral anterior nuclear rim and applied to its surface. Suction builds and the rim is drawn centrally. Brief taps of power draw the anterior and midlevel rim into the tip's aperture, fracturing it away from the deeper nuclear plate.

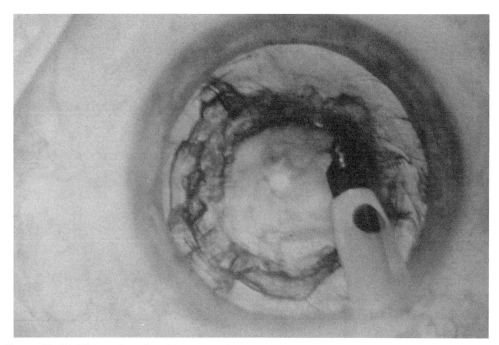

Figure 12-37. The nucleus has been rotated two or three clock hours counterclockwise with the cyclodialysis spatula. The deeper nuclear rim is shaved and thinned by the 45-degree tip in its cutting attitude.

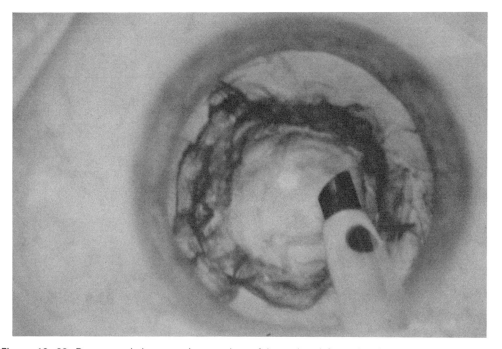

Figure 12-38. Deeper and deeper strips can be safely sculpted from the deeper nuclear layer in preparation for fracturing at these thin zones.

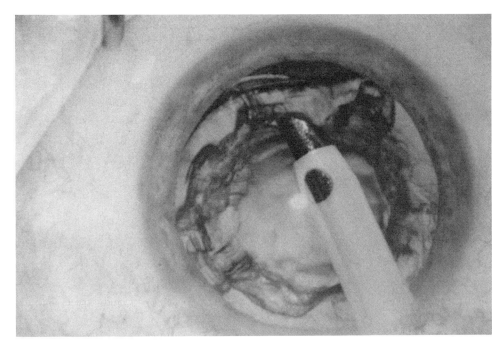

Figure 12–39. Peripheral anterior nucleus is drawn into the phaco tip in its suction mode.

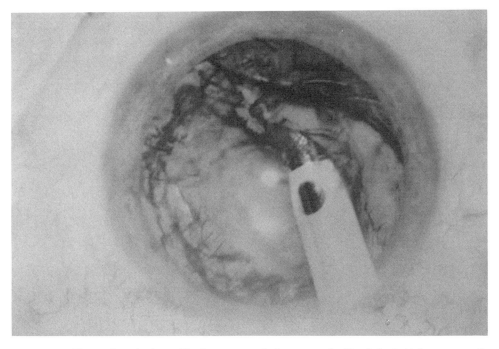

Figure 12–40. The nuclear rim is readily drawn centrally for removal with relatively high power suction and low power phacoemulsification.

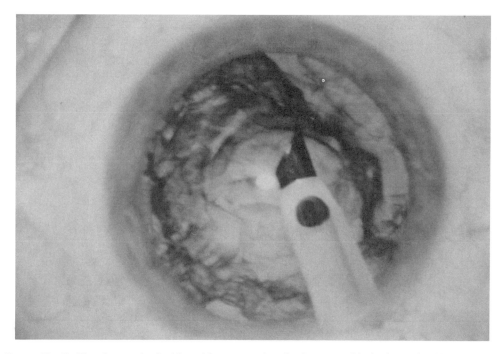

Figure 12–41. The deeper rim is thinned in preparation for fracture with the inward collapse of the anterior nuclear rim after one more rotation.

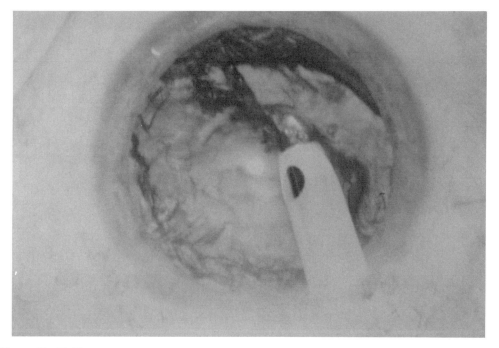

Figure 12–42. The anterior rim is drawn centrally.

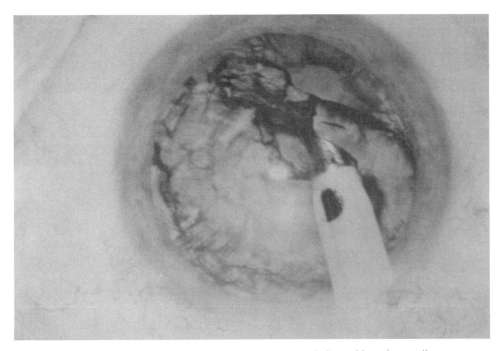

Figure 12–43. Portions of the rim are nibbled away while the tip is positioned centrally.

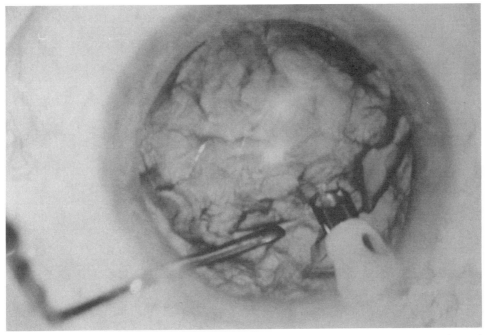

Figure 12–44. The posterior nuclear disc is now substantially reduced in bulk and diameter. It can be accessed with the cyclodialysis spatula and elevated slightly away from the posterior capsule before the tip is activated.

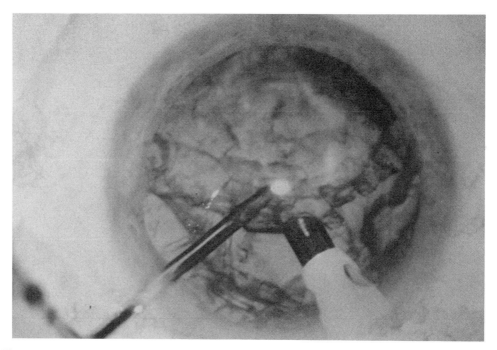

Figure 12–45. As the posterior disc becomes smaller, the phacoemulsification tip can be used a little more aggressively because the nuclear fragment can be held in a more central position away from the posterior capsule and visualization is improved.

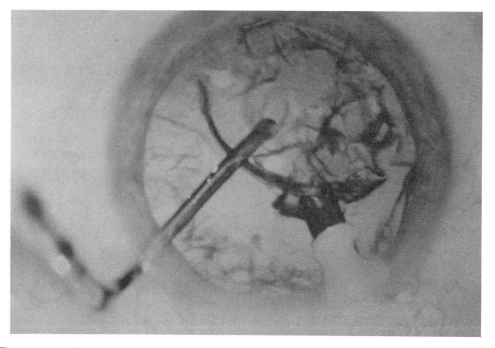

Figure 12–46. The spatula continues to support nuclear fragments away from the posterior capsule.

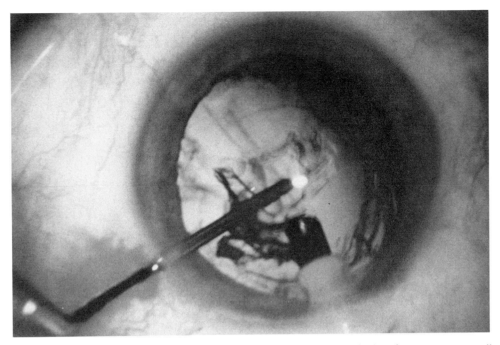

Figure 12–47. Phacoemulsification energy must be substantially reduced when fragments are small, or they will bounce off the tip when the power is activated.

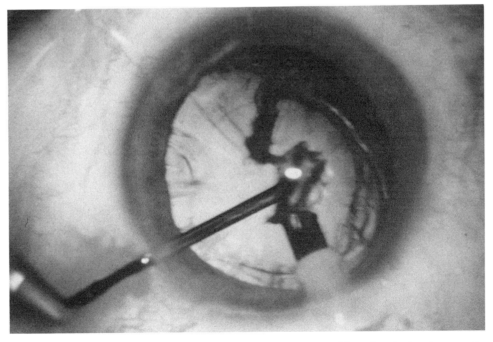

Figure 12–48. Watch for a flaccid posterior capsule that can be inhaled just after the last fragment has been aspirated. Keep the fragment away from the posterior capsule.

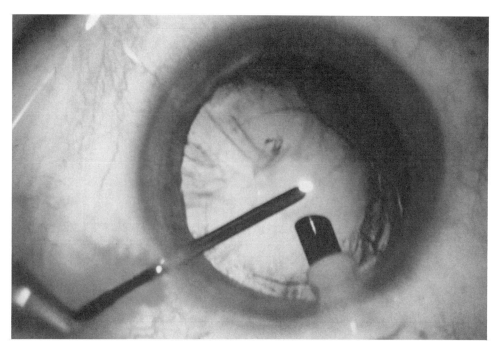

Figure 12–49. At the finish, the spatula should still be in position just deep to the level of the apex of the phaco tip, acting as an insulator between the tip and the posterior capsule.

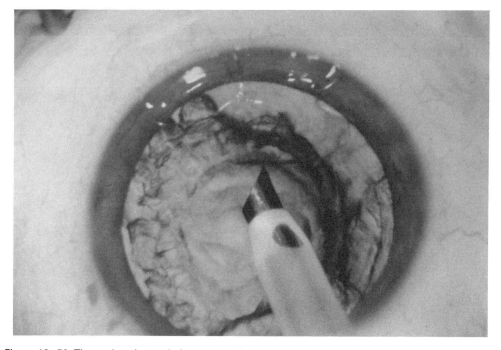

Figure 12–50. The nucleus is now in its new position with a new three clock hours available for inferior nuclear attack. The newly exposed midlevel nucleus is thin. The deeper peripheral nucleus is then visible for further thinning.

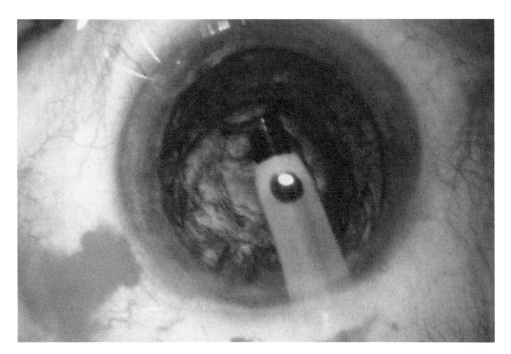

Figure 12–51. The phacoemulsification tip is prepared to sink into the remaining anterior peripheral nucleus and pull a wedge of it over the thinner, deep section thinned section. The deeper component is fractured so that the mid and anterior portions can be freed and aspirated.

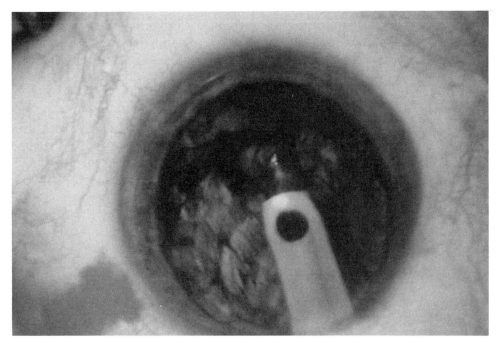

Figure 12–52. The aspiration of peripheral and anterior nuclear material is accomplished. After another rotation, the peripheral volume of the first 180 degrees has been reduced substantially so that aspiration of what is now inferior nucleus is easier and safer than the initial portions. More aggressive suction can be applied because the nuclear rim is able to be drawn centrally.

Cutting and Suction Technique for Soft Nuclei

The phacoemulsification procedure used in this technique has four basic steps:

(1) central sculpting
(2) debulking the inferior equatorial nucleus
(3) thinning, inward collapse, and aspiration of the remaining equatorial nucleus
(4) removal of the posterior nuclear disc.

1. Central Sculpting

With the infusion on, the 45–degree phacoemulsification tip is introduced with the bevel down so the tip does not traumatize the iris. After the bevel is turned up, sculpting begins. The tip should not be driven so deeply that the lumen becomes filled with nuclear material (Figure 12–24). During central sculpting, this material is shaved from the surface of the nucleus and the separated emulsate aspirated; i.e., the aspirated material is completely separated from the remaining nucleus.

It is easier to see and easier to avoid inadvertent occlusion by turning the tip on its side when emulsifying the sides of the peripheral nucleus (Figure 12–25). The tip should "bank" within the sulcus that is being created in the peripheral nuclear bowl much like a bobsled would carve through a turn in its track. The functional angle of attack of the 45–degree tip is shallow during only the first few passes. It becomes steeper with each pass.

Before removing much peripheral nucleus, the cyclodialysis spatula is used to confirm and complete the loosening of the nucleus from the adhesion of the cortex. The loosening can be completed by a posterior stirring kind of motion at three points of contact (Figures 12–26, 12–27, and 12–28). Be especially careful that the spatula does not sink through soft lenses.

As deeper nuclear material is emulsified, the functional angle of attack of the ultrasonic tip is gradually increased. With this increase comes an increased chance of inadvertent occlusion and posterior capsular damage. This is especially true if the tip is driven from superior to inferior along the posterior nuclear surface. The area just emulsified cannot be seen because it is hidden by the shaft of the ultrasonic tip. With the 45–degree tip, the aperture can be identified even at the steepest angle of attack required when thinning the posterior nucleus. This is especially critical in soft nuclei where softer more homogeneous material may actually seem to leap into the tip if it gets close to becoming occluded.

It is much safer and easier to turn the tip on its side and "sweep" the peripheral posterior nuclear surface (Figures 12–29, 12–30). This sweeping motion is effective for several reasons. First, the concavity of the posterior lens surface can be more accurately paralleled by the sweeping tip when pivoted at the wound than when the usual superior to inferior driving motion is used. The driving motion has a definite endpoint at the conclusion of each thrust.

This endpoint can be "overshot" by the surgeon, which gives the deeper nuclear drive a relatively high–risk conclusion. Second, tip occlusion is virtually impossible since the tip's aperture is always pointed toward the more open cavity of the nuclear bowl. Third, visualization of what the tip is actually doing is excellent. That is, the depth of the remaining nucleus can be seen immediately ahead of and behind the tip, showing the surgeon how much material is being shaved during the pass itself. Instantaneous adjustments in depth can be made and, because of this, the posterior peripheral nucleus can be shaved quite thin. Fourth, the nuclear material immediately under the anterior capsular remnant can be emulsified without danger of aspiration. This enables the surgeon to see the deeper, more peripheral material so he or she can carve further peripherally and posteriorly.

2. Debulking the Inferior Equatorial Nucleus

The surgeon's mode of attack now changes significantly. This is a transitional phase that occurs between the use of the cutting and suction modes primarily. It actually involves both, but in very brief, low power bursts.

As a safety measure, the maximum attainable power during linear control phacoemulsification is reduced to 60%. This is a maximum setting. In fact, levels of 10% to 30% are most often displayed on the control panel during this phase. The vacuum level stays increased to 61 mm Hg and flow rate reduced to 19 cc per minute when using the CooperVision Series Ten Thousand machines. Vacuum is increased to 33 inches of water but no alterations are made in the aspiration rate setting on the CooperVision 9001 series machines. Use a vacuum level of 58 mm Hg and a flow rate of 20 cc per minute with the Site peristaltic machine and a vacuum level of four inches with the Site diaphragm machine.

The tip's aperture is driven slightly into the anterior peripheral nuclear remnant at the 3:30 o'clock position so that the tip becomes occluded. Very brief taps of the foot pedal in the lowest phacoemulsification energy possible are used to withdraw and cut off a small section of peripheral nucleus (Figures 12–31, 12–32). This piece does not have to be deep or large because each successive section can be a little deeper and larger. In this way, the tip never becomes completely occluded by material that is directly attached to both sides of the remaining peripheral nuclear bulk except for the first small initiating section (Figures 12–33, 12–34, and 12–35). After three or four pieces have been removed, more underlying deeper newly exposed nuclear material can be shaved away with the tip on its side.

3. Posterior Nuclear Thinning, Inward Collapse, and Aspiration of the Peripheral Nucleus

The transitional phase completed, the ultrasonic tip will now be operated at extremely low power, alternating between suction and cutting modes. First, the aperture of the tip is applied like a suction cup to the 8 o'clock peripheral nuclear remnant. As suction is applied across the aperture, this one or two clock–hour section of peripheral nucleus folds in centrally as the deeper fibers fracture. The folding and fracturing actions usually occur just central to the posterior equatorial zone where the posterior nucleus has been thinned the most during the cutting mode. Brief taps of very low ultrasonic energy can be applied to encourage suction, infolding, and aspiration (Figure 12–36). It is important not to pursue peripheral nuclear material that does not come easily. The idea is to only debulk the peripheral nucleus by aspirating the anterior and equatorial level material while letting the posterior layer drop back after the fracture and separation of the posterior fibers.

The lens nucleus is then rotated counterclockwise approximately three clock hours. This can usually be accomplished with the cyclodialysis spatula alone but may require a bimanual technique using the emulsification tip as well. This exposes another three clock hours of the peripheral anterior and equatorial nucleus to aspirate with the assistance of low power ultrasonic energy. The tip is first returned to a cutting-shaving attitude to thin the deeper nuclear layers (Figures 12–37, 12–38). Again, only low power is used and great care is taken to avoid tip occlusion and burst through. The anterior nucleus will fold in and aspirate well if the deeper peripheral nucleus is adequately thinned so that the remaining deeper fibers will fracture. This deeper nuclear layer must fold and ultimately fractures, as the anterior and equatorial layers are aspirated with suction and brief taps of low power phacoemulsification energy (Figures 12–39, 12–40). Approximately half of the peripheral nucleus has been debulked after this second three clock–hour section has been emulsified.

Another rotation is accomplished and the posterior nucleus thinned with the cutting action of the ultrasonic tip. After this, the tip is again rotated 180 degrees into a suction orientation to accomplish the desired removal of the anterior and equatorial nuclear remnant (Figures 12–41, 12–42, and 12–43).

With each rotation, the posterior, thinning, infolding and peripheral aspiration occur more easily because the tip can be drawn more centrally while emulsifying and aspirating peripheral nucleus in the suction mode. This more central location is achieved because of the diminished nuclear bulk 180 degrees away and the resultant smaller nuclear diameter. Longer duration (but still low power) taps of emulsification energy can be used to emulsify and aspirate the infolded peripheral nucleus as it fractures. If too much material is drawn into the tip or if the nuclear shell wants to come forward and endanger the perfection of the anterior capsular remnant, it can be pushed away by the spatula.

4. Removal of the Posterior Nuclear Disc

The cyclodialysis spatula can be maneuvered so it can lift the reduced diameter posterior nuclear disc away from the posterior capsule (Figure 12–44). The superior pole of the posterior nuclear disc can be elevated slightly by the cyclodialysis spatula and guided into position for safe ultrasonic tip exposure.

Ultrasonic energy and suction can then be safely applied with the spatula always supporting the material to protect the posterior capsule (Figure 12–45). The shallow functional angle of attack of the 45-degree ultrasonic tip helps prevent posterior capsule aspiration. The spatula supports the nuclear fragments well away from the posterior capsule, which, if flaccid, can be easily aspirated at this point if the tip were deep (Figure 12–46). Remember that in most machines the vacuum and flow rate parameters are not varied by foot position and are just the same as when starting emulsification. After the last fragments are drawn to the tip aperture by suction, ultrasonic energy has to be very low or the fragments will merely bounce off the tip and never be aspirated. The cyclodialysis spatula helps hold the material in place while protecting the posterior capsule (Figures 12–47, 12–48, and 12–49).

Multiple Wedge Technique for Meduim Soft Nuclei

This technique works well for either soft or fairly firm lenses and is merely a variation of the suction-cutting method. It is a good technique to use if an individual nucleus seems too hard for use of the suction cutting technique but the surgeon is uncomfortable with the posterior cracking maneuver of Gimbel and Shepherd or if he has sculpted so far that there is not enough nuclear tissue left to convert to the cracking maneuver.

Initial sculpting and debulking of the first three inferior clock hours of peripheral nucleus are the same as the cutting suction technique for soft lenses. The nucleus is then rotated three clock hours and the newly presented posterior inferior nucleus thinned in the cutting mode (Figure 12–50). Three or four more wedges of nuclear rim are then emulsified and aspirated as they are cut away just as the initial three had been (Figures 12–51, 12–52). After additional rotations, the posterior nuclear disc is removed just as in the cutting-suction technique.

Posterior Nuclear Fracture and Quartering Technique for Firm Lenses

This method is one of the cracking procedures and is discussed in Chapter 16.

Section VI

ENDOCAPSULAR PHACOEMULSIFICATION WITH NUCLEAR CLEAVAGE

13

Two-Handed Phacoemulsification Through a Small Circular Capsulorhexis

I. Howard Fine, MD

The "Chip and Flip" Technique

Many, perhaps most, cataract surgeons who routinely use phacoemulsification agree that the most controlled technique is two-handed phacoemulsification within the pupillary plane. This technique has distinct advantages for the beginning phaco surgeon in the transition from extracapsular cataract surgery to phacoemulsification, when employing a very large anterior capsulotomy to facilitate prolapse of the superior equator of the nucleus into the plane of the pupil.

Recently, following the leads of Drs. Thomas Neuhann in West Germany and Howard Gimbel in Canada, it has become increasingly apparent that the continuous tear circular capsulorhexis with a wide anterior capsular flap has definite advantages. These include avoiding extensions of anterior capsule tears to the posterior capsule and ensuring in-the-bag placement and centration of posterior chamber implants. There are also theoretical advantages with respect to maintaining the sequestration of the anterior segment from the posterior segment following a posterior YAG laser capsulotomy.

The utilization of a circular capsulorhexis with a wide anterior capsular flap presents new challenges to phaco surgeons, the most important of which is an inability to easily dislocate the superior pole of the nucleus for pupillary plane phacoemulsification. Many have returned to the use of one-handed techniques they had previously abandoned in favor of two-handed techniques, which they felt offered more control and safety.

The Chip and Flip technique increases safety and control for phacoemulsification within the capsular bag. Surgical maneuvers that use a two-handed technique through a small circular capsulorhexis are described here.

Employing a blunt dissection technique, a Beaver blade is used to make a groove, and the same blade is used to dissect a scleral tunnel. The side port is made to the left with an especially sharp instrument, and viscoelastic is used to replace aqueous humor. The keratome is then used to enter the anterior chamber through the scleral tunnel. Using a bent needle, the circular capsulorhexis is started by making a small cut at the center of the lens, pulling directly toward the paracentesis and then curving toward the left. This creates a central flap that tears in a circular pattern to the right. The flap is folded over, purchased with Kraff-Utrata forceps, and pulled by the forceps in a circular motion so that the force at that point of tear is tangential to the circumference of the circle. It is necessary to repurchase the capsular flap closer to the tear point during the course of the capsulotomy. The completion of the tear comes from slightly outside toward inside so that a nick in the capsular ring does not result.

Hydrodissection is performed using the technique taught me by Dr. Thomas Neuhann of West Germany. A 26-gauge cannula is directed toward the center of the nucleus as deeply as one can go, and balanced salt solution is injected tangentially, resulting in the dissection of hard, central and soft, outer nuclear zones. A second hydrodissecting fluid wave may be placed just under the anterior capsule to facilitate rotation of the lens within the capsule. The phacoemulsification procedure then takes place through the small circular capsulorhexis. Central sculpting is accomplished in the usual manner. The Bechert nucleus rotator is brought in and *the nucleus is pushed toward 12 o'clock with the Bechert nucleus rotator under the tip of the phaco handpiece.* The rim of the inner nuclear bowl is removed at 5 o'clock to 6 o'clock; the nucleus is rotated clockwise, and another hour of rim is removed at 5 o'clock to 6 o'clock. Pulsed phacoemulsification adds an additional measure of control to avoid breaking through to the outer nuclear rim while segmentally removing the entire rim of the "inner" nuclear bowl.

There is usually a clear-cut demarcation line between the hard inner nuclear bowl and the soft outer nuclear bowl. *The displacement of the nucleus toward 12 o'clock with the emulsification taking place in the 5 o'clock to 6 o'clock zone protects the capsule* because the part of the nucleus being emulsified is brought away from the capsular fornix and out from under the iris, even in the presence of a small pupil. Emulsification takes place just under the anterior capsular flap and close to the center of the deepest place in the anterior chamber.

Once the rim of the inner nuclear bowl has been removed, the second handpiece can be brought into the cleavage plane between the inner nuclear

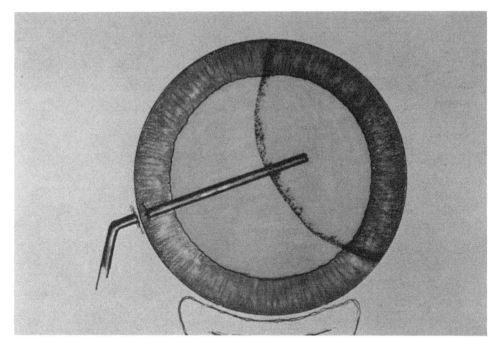

Figure 13–1. Viscoelastic is used to replace the aqueous humor.

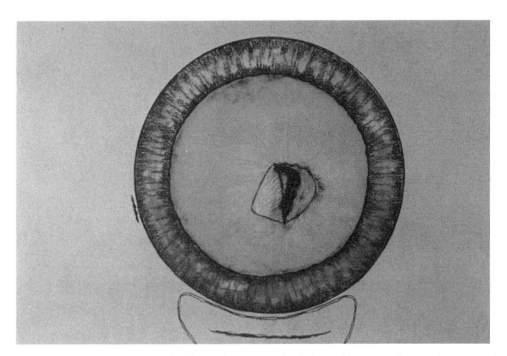

Figure 13–2. A bent needle is used to begin the capsulorhexis by making a small cut at the center of the lens, pulling toward the paracentesis and then curving to the left.

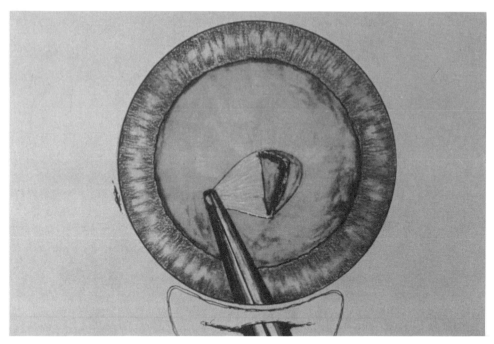

Figure 13–3. Utrata-Kraff forceps are used to grasp the central flap and pull it in a circular motion.

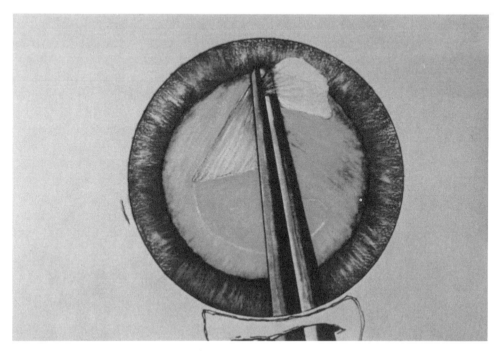

Figure 13–4. The tear is made so that the force at the point of the tear is tangential to the circumference of the circle.

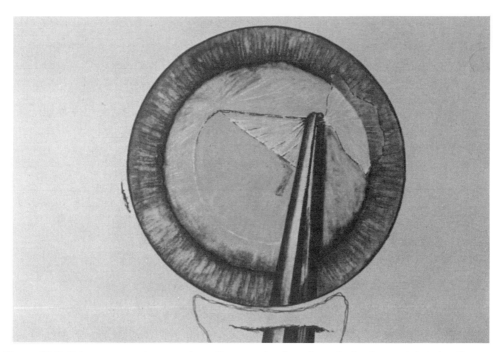

Figure 13–5. It is necessary to repurchase the capsular flap closer to the tear point during the capsulotomy.

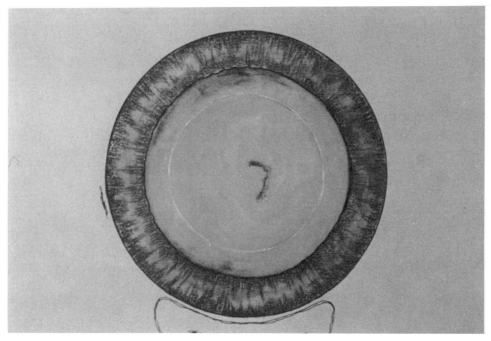

Figure 13–6. Completion of the tear comes from slightly outside toward inside so that a nick in the capsular ring does not result.

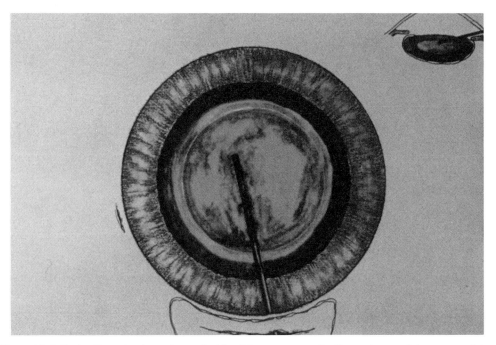

Figure 13–7. Hydrodissection is performed with a 26–gauge cannula directed toward the center of the nucleus as deeply as one can go, and balanced salt solution is injected.

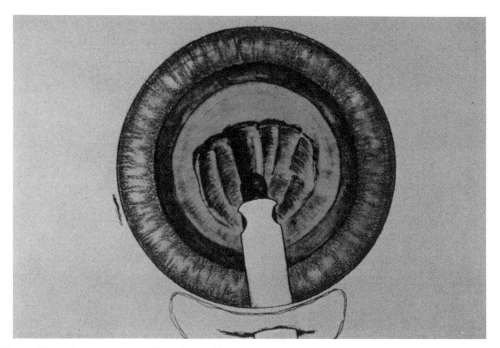

Figure 13–-8. Central sculpting is performed in the usual manner.

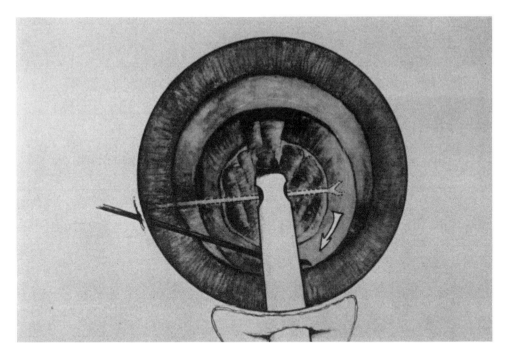

Figure 13–9. The rim of the inner nuclear bowl is removed at 5 o'clock to 6 o'clock.

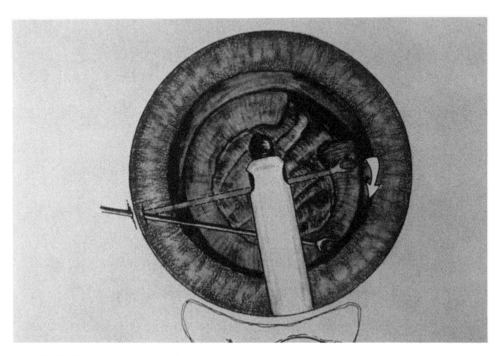

Figure 13–10. The nucleus is rotated, and another piece of nucleus is removed at 5 o'clock to 6 o'clock.

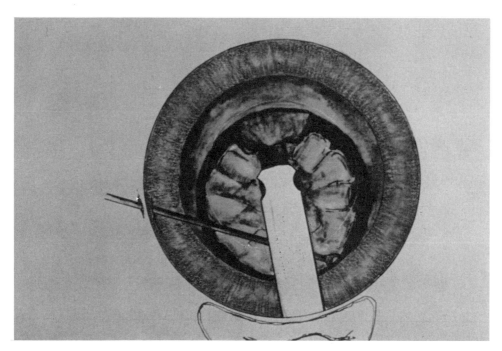

Figure 13–11. The nucleus is displaced toward 12 o'clock to facilitate removal of the rim of the inner nuclear bowl.

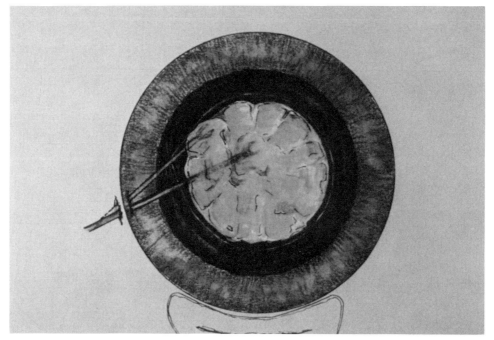

Figure 13–12. The second instrument is passed in the cleavage plane between the inner nuclear chip and the outer nuclear bowl, lifting the chip into the center of the capsular bag.

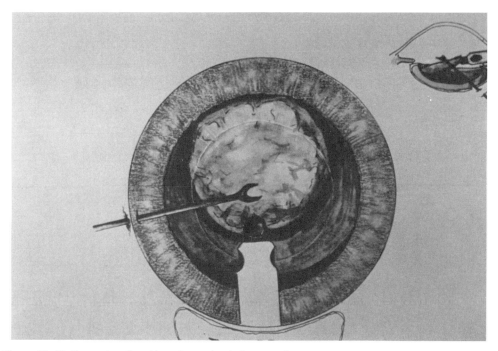

Figure 13–13. Removing the chip using pulsed ultrasound.

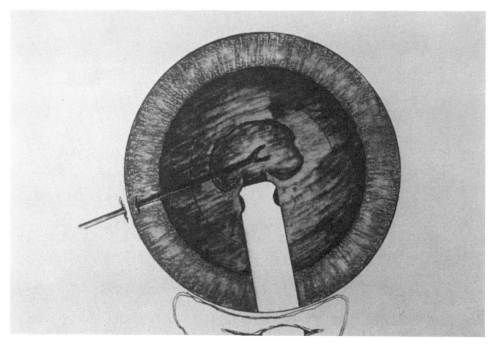

Figure 13–14. Removal of the last piece of inner nuclear chip.

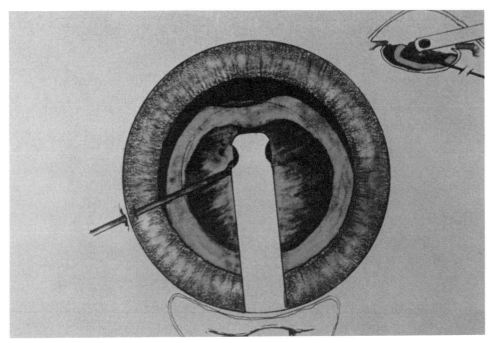

Figure 13–15. The outer nuclear bowl is displaced from the capsular fornix at 5 o'clock to 6 o'clock.

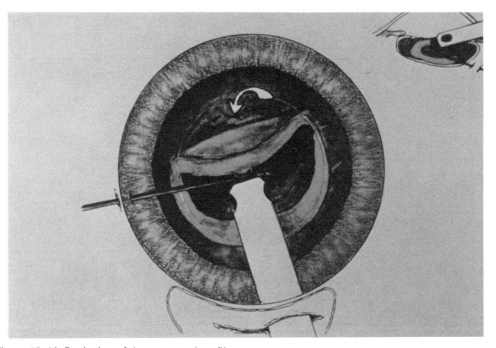

Figure 13–16. Beginning of the outer nuclear flip.

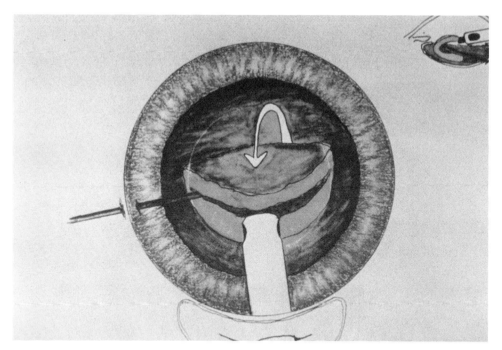

Figure 13–17. The second instrument pushes the bottom of the outer nuclear bowl toward 5 o'clock to 6 o'clock.

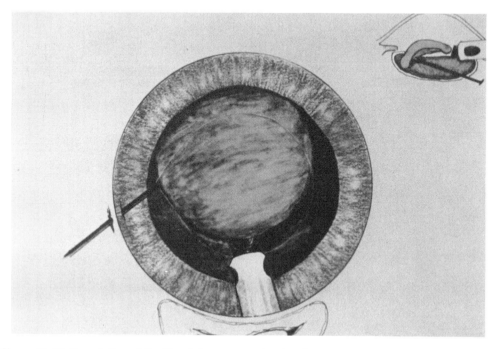

Figure 13–18. Completion of the flip of the outer nuclear bowl.

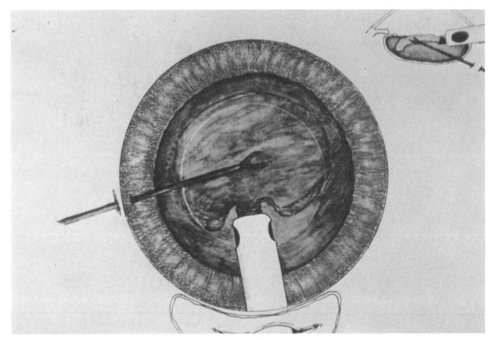

Figure 13–19. The outer nuclear bowl is easily removed using aspiration or low levels of emulsification.

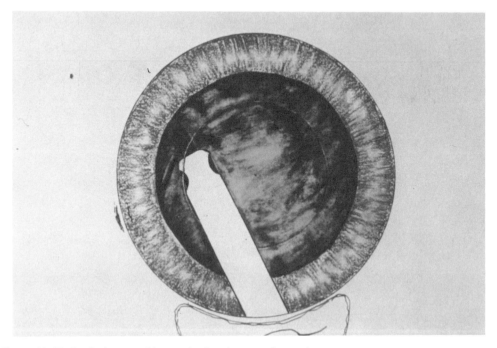

Figure 13–20. Cortical removal is easy in the absence of capsular tags.

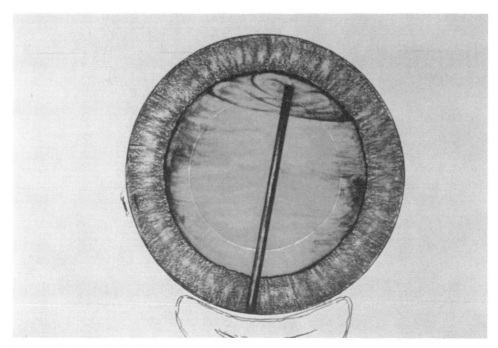

Figure 13–21. Viscoelastic fills the capsular bag.

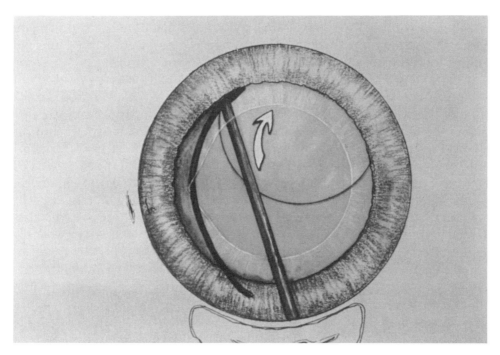

Figure 13–22. The implant is directed under the anterior capsular flap and rotated into position.

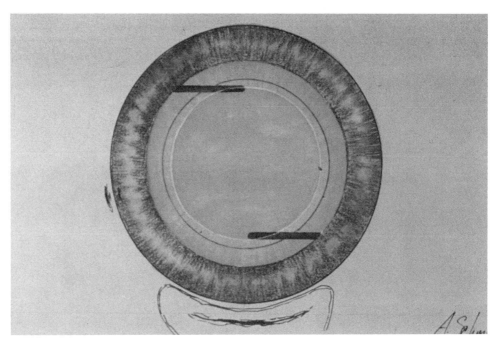

Figure 13–23. Visual confirmation of implant position and centration.

chip and the outer nuclear bowl and swept under the chip, elevating it into the center of the capsular bag. Using the second handpiece to control the nuclear chip, the chip can be quickly and safely removed. Pulsed phacoemulsification dramatically reduces "chattering" of the chip.

The soft outer nuclear bowl, which has cushioned all previous phacoemulsification, is now displaced from the capsular fornix at 5 o'clock to 6 o'clock. This is done either by pushing the 12 o'clock "rim" toward 12 o'clock, thus bringing the 5- to 6 o'clock rim out of the fornix, or by pushing the center of the bowl "toward" 5- to 6 o'clock, causing the 5- to 6 o'clock rim to curl out of the fornix and under the anterior capsulotomy flap back toward 12 o'clock. Using the phaco tip, in aspiration mode only, the nuclear bowl is mobilized by pulling the rim at 5 o'clock to 6 o'clock toward 12 o'clock and pushing with the second handpiece in the bottom of the nuclear bowl toward 5 o'clock to 6 o'clock to tumble or flip the soft outer nuclear bowl. Several attempts at this maneuver, with rotation of the bowl following each attempt, may be necessary in order to achieve the tumbling. By flipping the bowl away from the capsule, it can be removed safely either with aspiration or low powers of emulsification without jeopardizing the capsule.

The cortex strips easily and safely in the absence of capsular tags. The capsular bag is filled with viscoelastic. The implant is directed under the anterior capsular flap, and the optic is placed at least half way through the anterior capsulorhexis. The trailing loop can be dialed in, and one can see the distortion in the anterior capsular ring to the left as it peaks toward the pupil and then snaps back when the haptic goes under the anterior capsular flap. One can visually confirm that the implant is placed within the capsular bag and that centration is excellent.

14

Spring Surgery

Paul S. Koch, MD

Spring Surgery is a particular form of endocapsular phacoemulsification that takes advantage of the natural spring-like forces in the eye to provide a safe and effective way to remove the inner nucleus. The word "spring" is also an acronym for the *s*equential *p*ulsed *r*emoval of the *in*ner *n*uclear *g*irdle.

It is important to recognize from the outset that the nucleus is not a single homogeneous mass but consists of at least two specific structures, the inner nucleus and the outer nucleus (Figure 14–1). In a way, the inner nucleus corresponds to the fetal nucleus. It is firm and hard and relatively rigid. It is the most difficult structure to emulsify. When partially emulsified, its sharp edges are responsible for many of the problems that occur during phaco-emulsification, including tearing of the posterior capsule.

The outer nucleus, on the other hand, is a soft structure, and it can be relatively thick. If the outer nucleus can be isolated from the inner nucleus, it will serve as a sort of foam rubber cushion to protect the posterior capsule while we are working on the inner nucleus.

This is accomplished through hydrodelamination, in which a 30–gauge cannula is placed on a 3–cc syringe filled with BSS. This needle is placed in the eye to approximately the lateral cut edge of the anterior capsule and passed into the body of the nucleus until it meets resistance. This is usually about one-third of the way into the nuclear body. The resistance is actually the transition zone between the soft outer nucleus and the firm inner nucleus (Figures 14–2, 14–3).

Gentle and sometimes forceful irrigation is used at this point, and a cleavage plane can be hydrodelaminated in this fashion. A clear separation appears immediately between the inner and outer nucleus. In softer nuclei, this ring may completely surround the inner nucleus all at once. In other eyes it is necessary to pass the cannula into the body of the nucleus again, on the other side of the cataract, and re-inject at that point.

The goal of hydrodelamination is to create a cleavage plane between the inner nucleus and the outer nucleus (Figure 14–4). The location of the cleavage planes depends on how hard the nucleus is. In a soft cataract there will be a small inner nucleus and a large outer nucleus; in a firm cataract the inner nucleus will be larger and the outer nucleus will be smaller. In some very firm cataracts it will be nearly impossible to isolate an outer nucleus.

The next step of the preliminary maneuvers is hydrodissection, in which a cleavage plane is opened between the outer nucleus and the cortex. This is done with the same cannula, but this time it is placed between the cataract and the capsular bag and fluid is injected under the capsular bag. The fluid wave can be seen to pass right around the back of the cataract and up the other side again, separating the nucleus from the cortex. Note that we are not separating the cortex from the capsule because, if we were, there would be no need for cortical aspiration at the end of the case (Figures 14–5, 14–6).

There are two cleavage planes at this point. The capsule and cortex are separated from the outer nucleus by the hydrodissection cleavage plane, and the outer nucleus is separated from the inner nucleus by the hydrodelamination cleavage plane. This still leaves us with the problem of how to remove a 6 mm to 8 mm inner nucleus through a 4 mm to 5 mm capsular axis opening.

At this point I would like to define an anatomic entity that I call the inner nuclear girdle. In fact, it is exactly what other people call the inner nuclear bowl. The inner girdle is a ring much like the girdle of a tree. The outside edge of the ring is the hydrodelamination plane separating the inner nucleus from the outer nucleus. The inner boundaries of the ring is the space created when one sculpts out the inner nucleus (Figure 14–7).

If one looks at the cataract from above, as through the microscope, the inner nuclear girdle looks something like a donut. There is a space in the middle that was the sculpted area; then comes the mass that I call the inner nuclear girdle; and then there is a ring outside of that, which is the cleavage plane between the inner nucleus and the outer nucleus.

The cleavage planes we have formed in hydrodelamination and hydrodissection are natural cleavage planes made apparent by the forceful use of a fluid injection. It is also possible to obtain an autocleavage plane. We have seen this in one patient who was admitted to the hospital with acute hyperglycemia and a blood sugar greater than 600. The patient's complaint of blurred vision was assumed to be due to the hyperglycemia. However, it was discovered on examination that the hyperglycemia had caused a lot of fluid to be attracted to the lens and the lens underwent an autocleavage with a split right down its center (Figure 14–8).

The Machine in Spring Surgery

At this time it is necessary to be sure that the machine settings are appropriate for this task. While standard machine settings can be used for the sculpting phase of this procedure, once we get into the spring mode it is important to have low levels of emulsification that can be established using the linear mode of the machine and low aspiration. In a diaphragm machine, this can be as low as 2 inches of vacuum. In a peristaltic machine, it depends on the relationship between the vacuum level and the flow rate. A typical setting may be 58 mm on the vacuum and 20 cc on the flow rate or 80 mm on the vacuum and 15 cc on the flow rate.

The machine must also be set in the fast pulse mode. I prefer ten pulses per second. The pulse mode is valuable in providing a ratcheting effect so that the nucleus comes to the tip very slowly. Because of this, it is relatively easy to stop emulsification and aspiration at a predetermined point.

The Physiologic Trampoline

When the machine parameters are set, the spring mode begins. The phaco tip is placed against the inferior cut edge of the sculpted area and emulsification begins (Figure 14–9). The nucleus comes to the tip and is attracted to it. But if you hold the tip fixed at the inferior edge of the capsular axis, the nucleus will be attracted to the tip only to a certain point, and it will suddenly spring away from the tip without the surgeon doing anything.

Although it may look on videotape as though the surgeon suddenly let up on the emulsification, what really happens is that the nucleus, the capsular bag, and the zonules react like a physiologic trampoline that when stretched causes the nucleus to be propelled away and back to its normal central position.

Here is what happens. The nucleus is attracted to the phaco tip, and as it is drawn to the tip—remembering that the tip is being held in one place—the cleavage plane between the inner nucleus and the outer nucleus closes superiorly. As the nucleus continues to be attracted to the phaco tip and comes toward the tip, it will press on the superior portion of the bag and the cleavage plane between the outer nucleus, and the bag will close (Figure 14–10).

Once the inner nuclear girdle has moved in a superior direction as far as the tissues in the capsular bag will absorb, the bag itself is put on stretch and the inferior zonules become taut. This is the physiologic trampoline I mentioned earlier (Figure 14–11).

A trampoline is a stretched piece of fabric that is held securely to a frame by springs (Figure 14–12). If a little bit of tension is placed on the trampoline by someone jumping only a short distance, the springs stretch out only a little bit and the person on the trampoline is propelled just a short way into the air (Figures 14–13, 14–14). If a lot of tension is placed on the trampoline by someone jumping very high, the springs stretch out a lot and the person is propelled very high into the air (Figures 14–15, 14–16).

The same trampoline principle applies to the capsular bag. When the bag itself is stretched like the fabric of a trampoline and the inferior zonules are put on tension like the springs of a trampoline by the superior movement of the nucleus, it is inevitable that the nucleus will be propelled inferiorly as the physiologic trampoline reacts (Figure 14–17).

This is possible only if the phaco tip is on low levels of aspiration so that the nucleus is free to spring away. With higher levels of aspiration, the nucleus would be held there and damage could occur to either the capsular bag or the zonules.

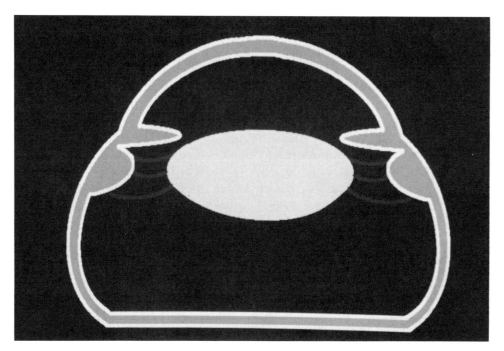

Figure 14–1. The nucleus is not a single homogeneous mass but consists of a firm inner nuclear and a soft outer nucleus.

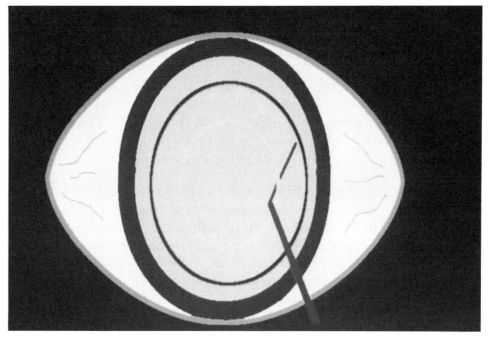

Figure 14–2. The inner nucleus is separated from the outer nucleus by hydrodelamination. A cannula is placed into the body of the nucleus, and balanced salt solution is injected into it.

Figure 14–3. The fluid separates a natural cleavage plane between the inner nucleus and the outer nucleus.

Figure 14–4. The inner nucleus is a firm structure, which is the most difficult to emulsify. The outer nucleus is a soft structure and serves as a cushion between the firm inner nucleus and the capsular bag.

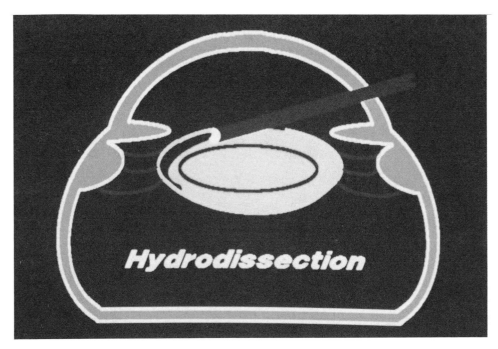

Figure 14–5. The nucleus is separated from the cortex by hydrodissection. Fluid is injected between the nucleus and the capsular bag, creating a cleavage plane between the outer nucleus and the cortex.

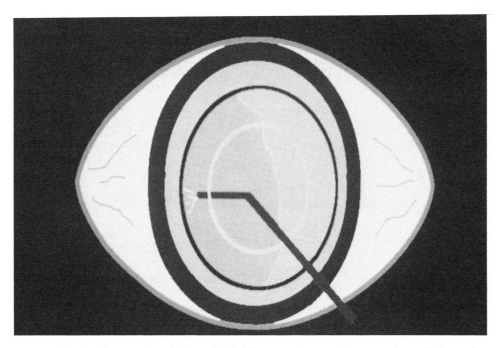

Figure 14–6. As the cleavage plane is formed, a fluid wave can be seen to pass underneath the nucleus.

Figure 14–7. The inner nuclear girdle is the round ring of inner nucleus. The inner border is the cut edge of the sculpted tissue, and the outer border is the cleavage plane between the inner nucleus and the outer nucleus.

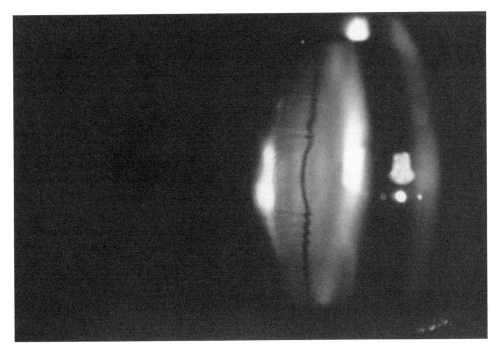

Figure 14–8. A case of autonuclear cleavage in a patient with acute hyperglycemia greater than 600 mg. A case of spontaneous cleavage of the nucleus in a patient with acute hyperglycemia. The blood sugar was greater than 600.

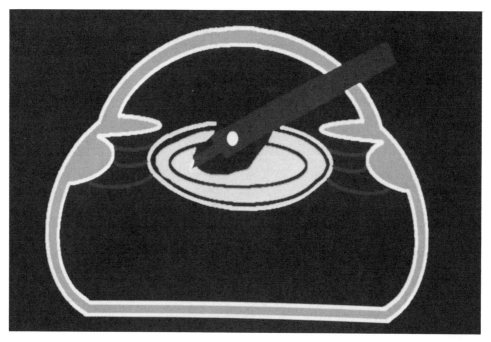

Figure 14–9. After sculpting, the phaco tip is placed against the inferior nucleus and is held steady while the nucleus is attracted to it.

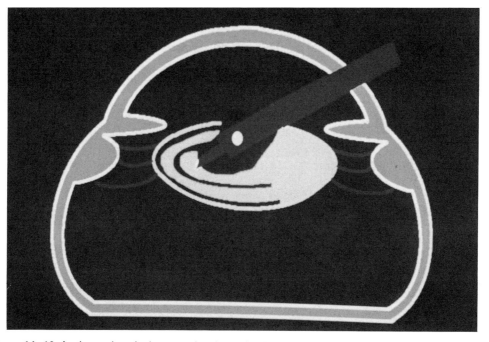

Figure 14–10. As the nucleus is drawn to the phaco tip, the cleavage plane between the inner nucleus and the outer nucleus closes, and the cleavage plane between the outer nucleus and the cortex closes.

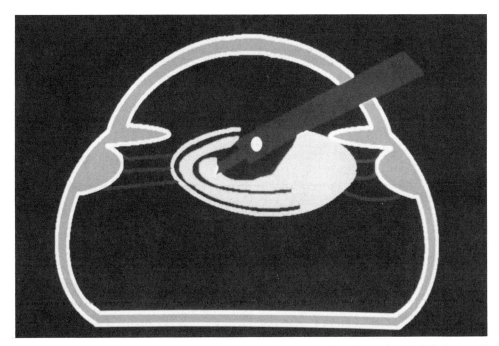

Figure 14–11. As the inner nucleus continues to be attracted to the phaco tip, the inferior zonules became taut.

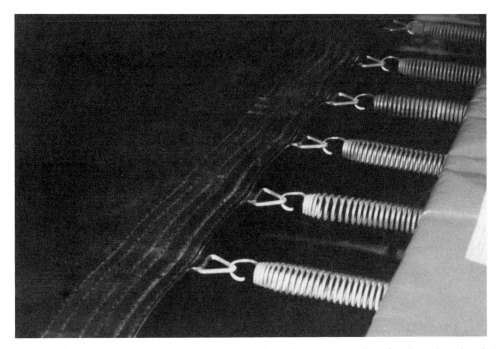

Figure 14–12. A trampoline is a piece of fabric held to its frame by a series of springs. Imagine that the posterior capsule is the fabric and the zonules are the springs.

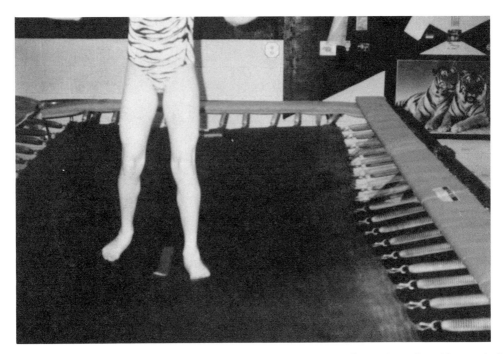

Figure 14–13. A little bit of tension is placed on the fabric (capsular bag), and the springs (the zonules) become taut.

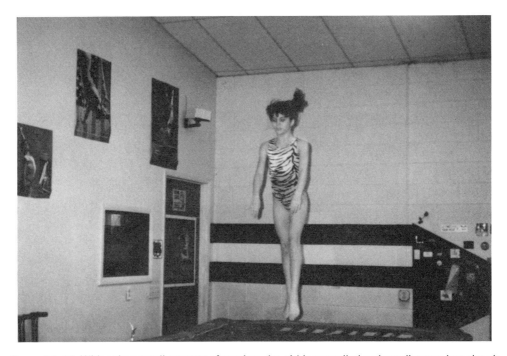

Figure 14–14. With only a small amount of tension, the girl is propelled a short distance into the air.

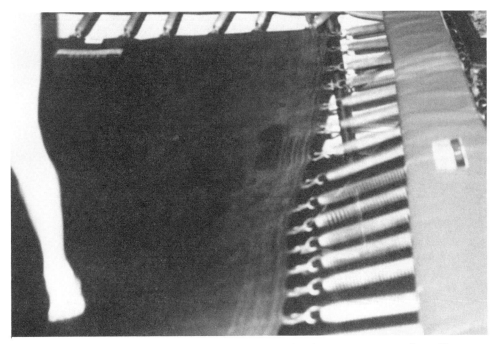

Figure 14–15. If a lot of tension is placed on the fabric and springs, a greater reaction will occur.

Figure 14–16. When the fabric and springs recoil, the girl is propelled with a good amount of force well into the air. Similarly, once the capsule and zonules become taut, the inner nucleus will spring back to its normal physiologic position.

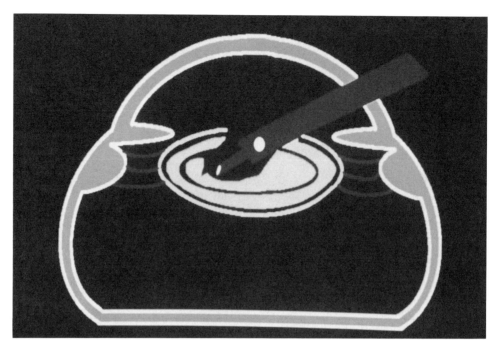

Figure 14–17. Continuation of surgical sequence seen in Figure 14–11. The nucleus has "sprung away" from the phaco tip and has resumed its normal physiologic position.

Figure 14–18. After half of the nucleus has been partially emulsified, thin inner nucleus is rotated superiorly, and the phaco tip begins to emulsify the new inferior nucleus.

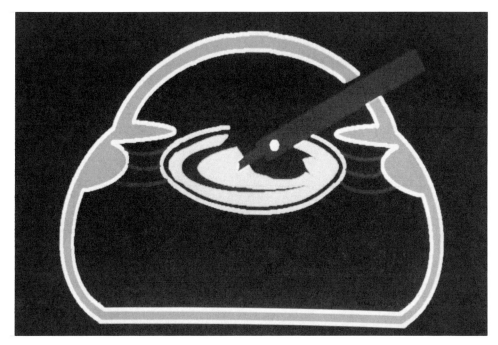

Figure 14–19. Because the superior nucleus has already been thinned, it gives way at the same time that the cleavage planes are being closed. This permits more inferior nucleus to come to the tip.

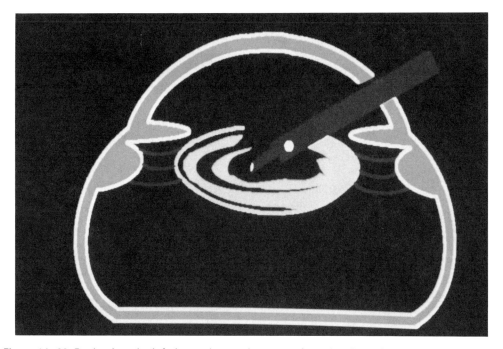

Figure 14–20. By the time the inferior nucleus springs away from the phaco tip, a lot more tissue has been removed than in the original passes earlier in the operation.

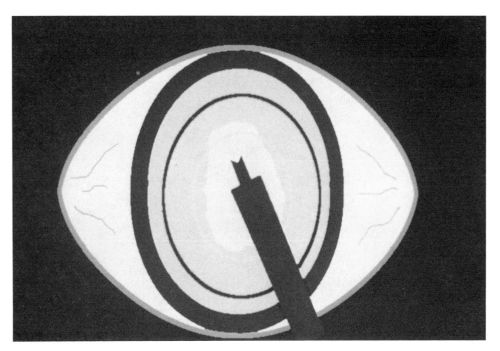

Figure 14–21. After central sculpting, the nucleus can move only a little bit superiorly, and only a little bit of inferior nucleus can be emulsified.

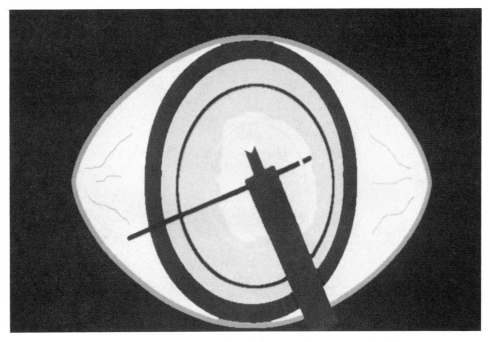

Figure 14–22. After the original spring away of the inferior nucleus, a spatula is introduced and the inner nuclear girdle is rotated a few clock hours.

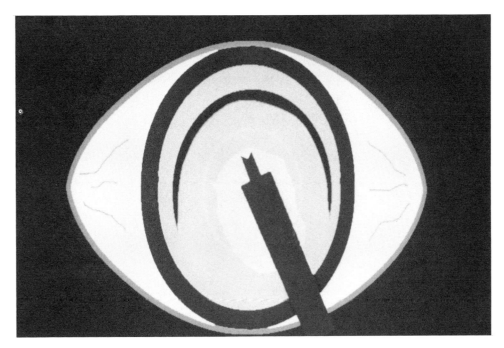

Figure 14–23. Additional thinning of the inner nuclear girdle is performed and, as the nucleus thins, the inferior cleavage plane opens wider.

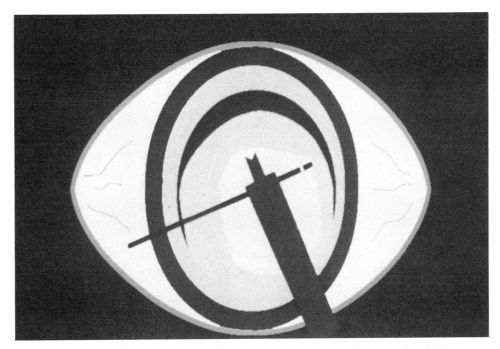

Figure 14–24. With each subsequent rotation, more and more tissue is removed and the amount of nucleus coming to the tip gets progressively greater.

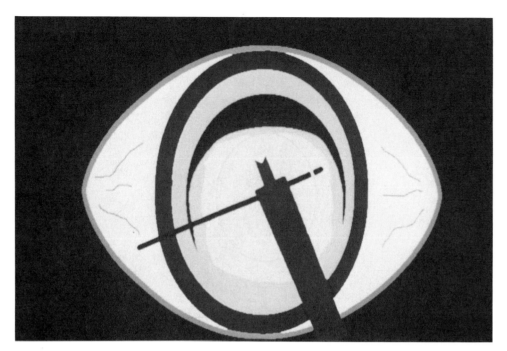

Figure 14-25. After one or two rotations, the inferior portion of the inner nuclear girdle is very thin and it is very easy to break through the wall.

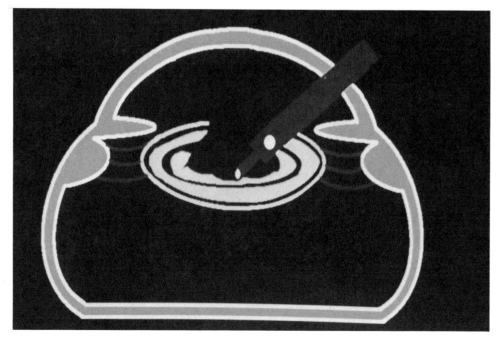

Figure 14-26. While the wall of the inner nuclear girdle is being thinned, the floor is also shaved so that the posterior plate will end up being thin and pliable.

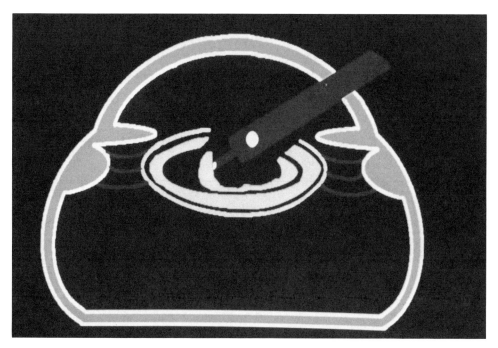

Figure 14–27. A final peripheral portion of the inner nuclear girdle is aspirated and, if the plate is thin, it will fold up and come up directly to the phaco tip.

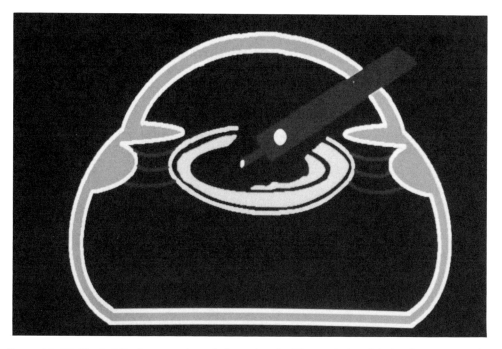

Figure 14–28. Most of the inner nuclear girdle and part of the posterior plate has been removed.

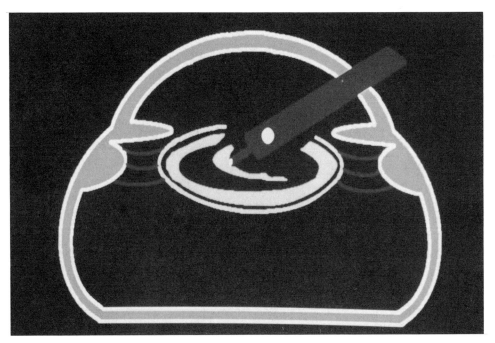

Figure 14–29. The remainder of the posterior plate is rotated into position and is easily aspirated well within the cushion of the outer nucleus.

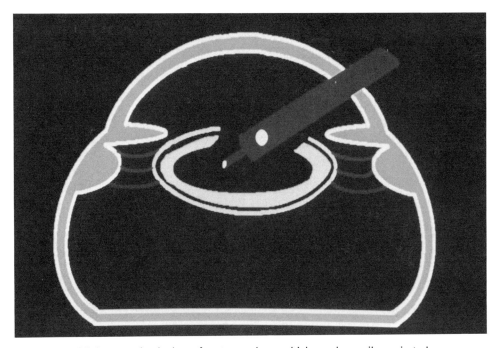

Figure 14–30. All that remains is the soft outer nucleus, which can be easily aspirated.

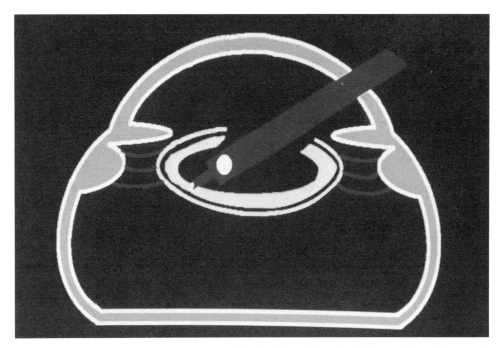

Figure 14–31. The outer nucleus is aspirated with the phaco tip and pulled away from the capsular bag.

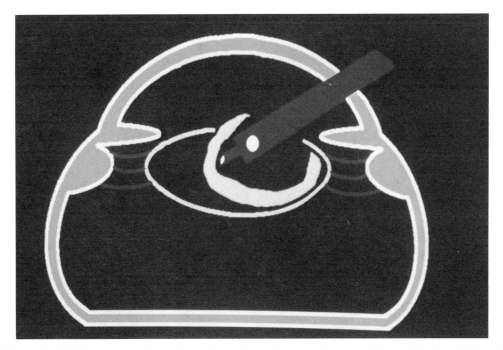

Figure 14–32. The outer nucleus is teased away from the capsular bag and brought to the middle of the bag for easy aspiration and emulsification.

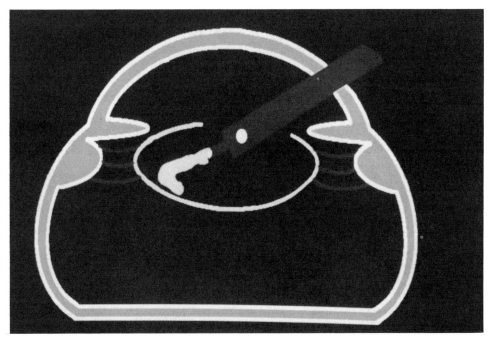

Figure 14–33. The outer nucleus frequently flips over and follows easily into the phaco tip.

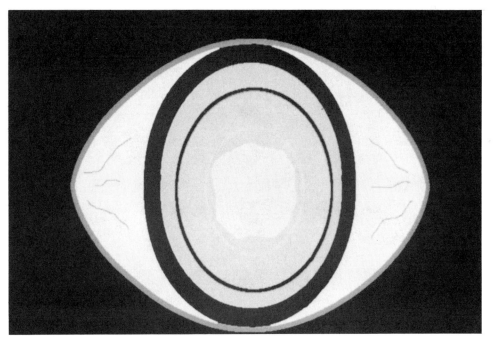

Figure 14–34. The hinge method also begins with sculpting of the inner nucleus.

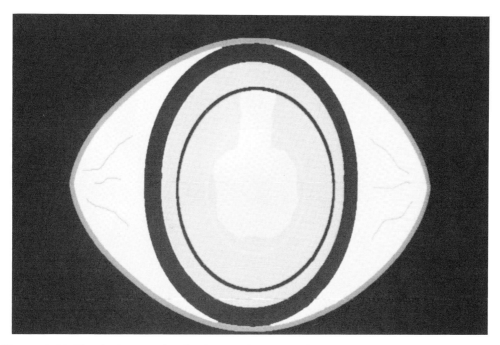

Figure 14–35. After the inner nucleus has been sculpted, the inferior rim is emulsified repeatedly using the principles of Spring Surgery until the rim of the inner nuclear girdle is broken, creating a free communication between the sculpted area and the cleavage plane between the inner nucleus and the outer nucleus.

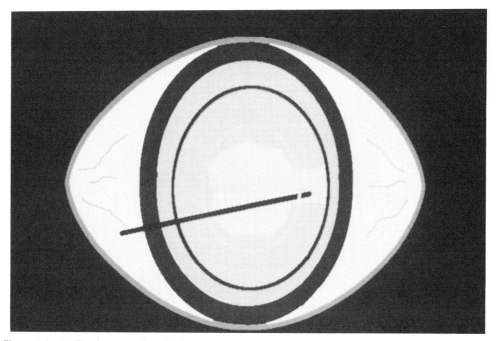

Figure 14–36. The inner nucleus is then rotated with the aid of a spatula.

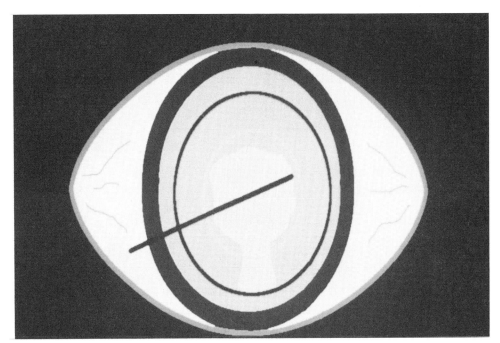

Figure 14–37. Rotation terminates when the previous cut edge of the inner nuclear girdle is at 12 o'clock.

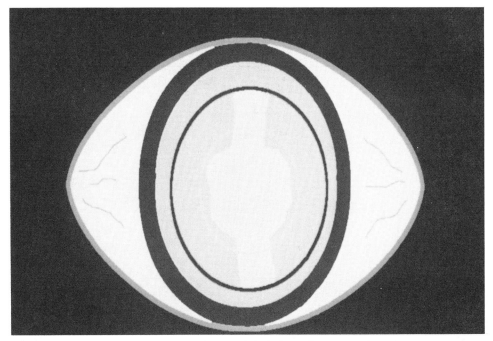

Figure 14–38. Additional sculpting is performed inferiorly until the rim of the inner nuclear girdle breaks so that the girdle itself has been bisected.

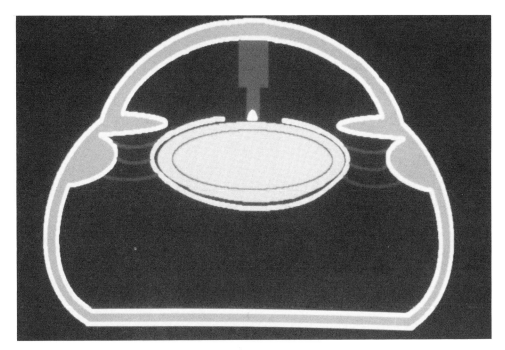

Figure 14–39. In cross section, the sculpting is performed in such a way that a trough will be created.

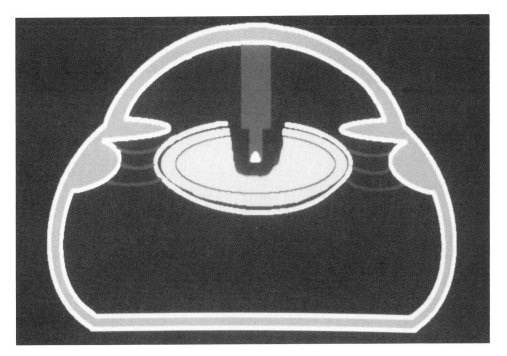

Figure 14–40. When the central sculpting has been completed and the inferior rim has been broken, sculpting continues to be directed posteriorly.

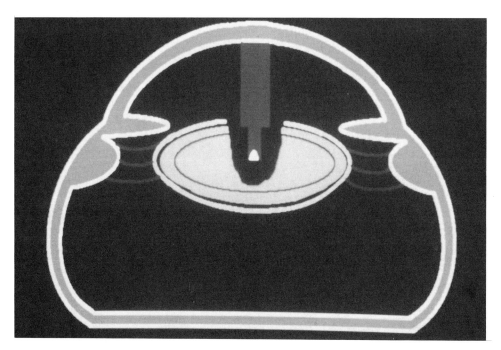

Figure 14–41. Continued sculpting of the inner nucleus, taking progressively deeper pieces of the nucleus.

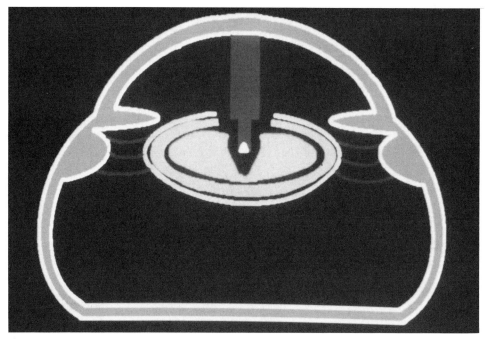

Figure 14–42. When the posterior plate of the nucleus is sufficiently thin, the two halves of the nucleus will "hinge" in toward the center of the capsular bag.

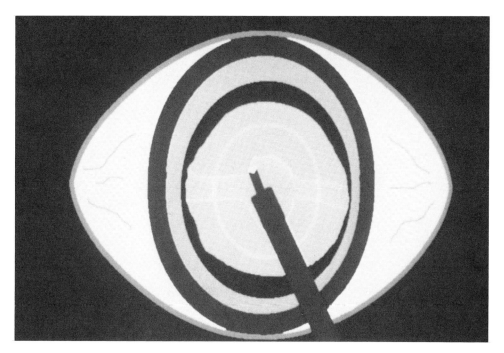

Figure 14–43. Once the two halves begin to hinge in, they are rotated 90 degrees, and each half is aspirated toward the middle of the capsular bag and emulsified there.

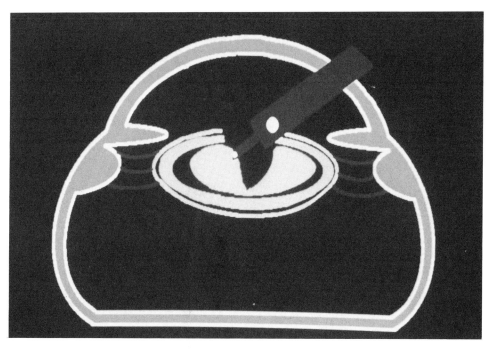

Figure 14–44. If the floor of the nucleus is very thin, half of the nucleus will come to the center of the phaco bag quite easily.

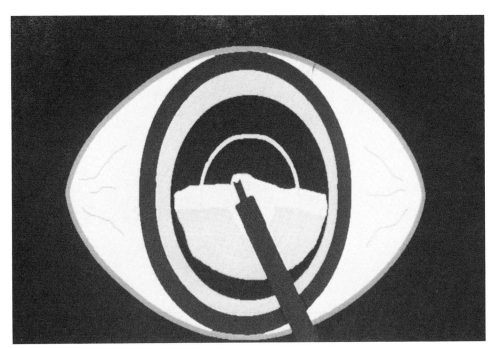

Figure 14–45. A cleavage plane will open between the inner nucleus and the outer nucleus so that only the inner nuclear half will be emulsified.

Figure 14–46. Emulsification of the first half of nucleus continues. If it is very firm, it may need to be quartered rather than halved.

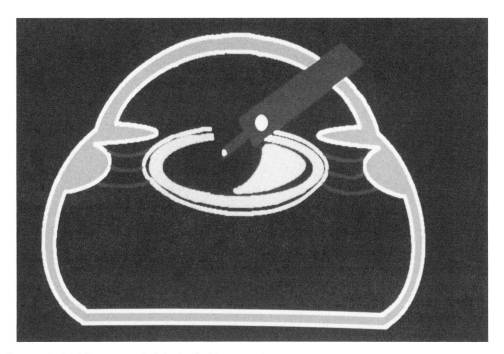

Figure 14–47. After removal of the half of inner nucleus.

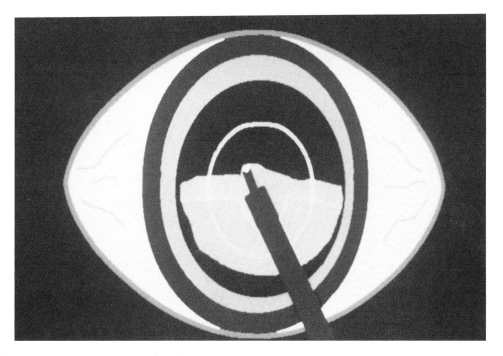

Figure 14–48. Once one half of the inner nucleus is removed, the second half needs to be rotated.

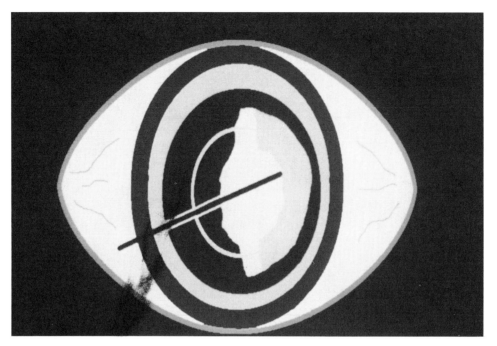

Figure 14–49. Rotation of the second half of the nucleus.

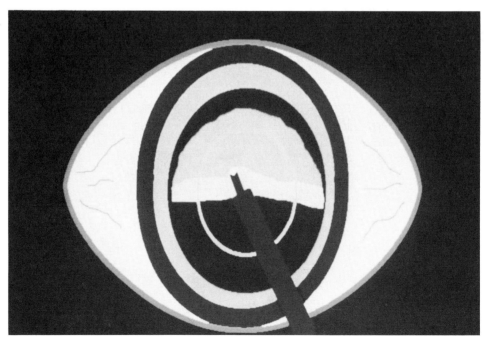

Figure 14–50. Aspiration and emulsification of the final half of the nucleus.

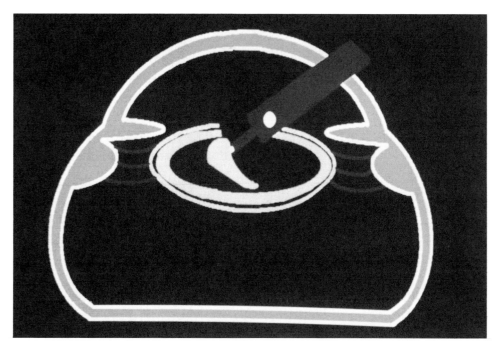

Figure 14-51. The final chip of nucleus is lifted off the outer nucleus and emulsified in the middle of the capsular bag.

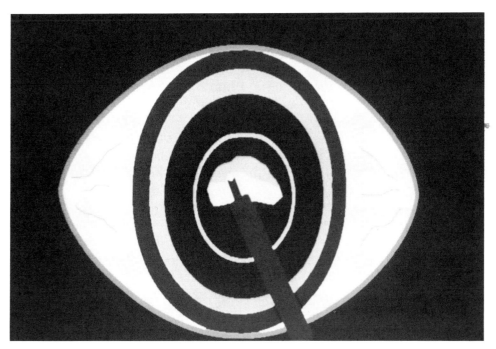

Figure 14-52. The final chip of posterior plate and nucleus is removed in the middle of the capsular bag, leaving the outer nucleus intact for aspiration and emulsification as before.

This is the key to Spring Surgery. By using low levels of aspiration and low phaco powers, it is possible to draw the nucleus to the phaco tip only until it activates the physiologic trampoline in the eye. The trampoline then propels the inner nuclear girdle back to its normal position, relieving all stresses on the capsular bag and zonules. This is the main safety feature of Spring Surgery.

The Rotation Method

There are two methods to Spring Surgery. The first is the rotation method, which works better in firmer nuclei. The second is the hinge method, which works somewhat better in soft nuclei.

The rotation method begins with deep central sculpting to about the limits of the capsular axis. Additional deep and wide sculpting can be performed as far as the surgeon feels comfortable.

Attention is then turned to the inferior portion of the sculpted zone to the inferior inner border of the inner nuclear girdle. This point will be aspirated to the tip and in a moment it will "spring away" from the tip.

After the nucleus springs away, it is rotated a clock hour to two and this maneuver is repeated. It can be repeated once or twice at each clock hour, trimming away the more anterior nucleus and then the more posterior portions of that same peripheral ring. If the floor of the nucleus requires thinning, it can be shaved down at the same time.

Thin, previously shaved tissue reaches the 12:00 o'clock position after about 180 degrees of this maneuver. This tissue has more give than the unshaved tissue that was there previously therefore, when one begins to emulsify at the 6:00 o'clock position and draw the nucleus superiorly, there is more give before the physiologic trampoline begins to stretch, and more tissue can be removed at 6:00 o'clock than previously. Likewise, once 360 degrees of inner nuclear girdle has been thinned, the tissue at 12:00 o'clock is thinner yet and progressively greater amounts of nucleus come to the tip inferiorly (Figures 14–18 through 14–25).

The inner nuclear girdle is eventually broken and totally separated, leaving only a very thin and small plate of posterior nucleus, which floats up easily and is removed without difficulty in the middle of the capsular bag (Figures 14–26 through 14–30).

This rotation technique of Spring Surgery is not unlike other things that have been discussed in this book. Dr. Davison's chapter on capsular bag phaco describes techniques for thinning the nuclear rim, but his description is in the case of a nucleus that has not had hydrodelamination. His techniques are also applicable to nuclei that have had hydrodelamination.

In the previous chapter of Dr. Fine's Chip and Flip maneuver, he uses a spatula to push the nucleus toward the 12:00 o'clock position, separating the 6:00 o'clock inner nucleus from the 6:00 o'clock outer nucleus and enlarging the cleavage plane there to make it easier to break through the inner nuclear wall.

The difference here in Spring Surgery is that we are concentrating on using the capsule and the zonules as a physiologic trampoline to propel the nucleus back to the middle of the eye so that we are always acutely aware that there are never any abnormal forces placed on the capsule or zonules. We never use a spatula or the strong force of the phaco tip to manipulate the nucleus, because those forces might overwhelm the capsule or zonules. Spring surgery is a more physiologic way to manipulate the nucleus.

Once the inner nucleus is gone, the outer nucleus can be aspirated with a phaco tip and massaged off the capsule and teased into the middle of the capsular bag, sometimes using the spatula to help flip it over—as Dr. Fine describes in the Chip and Flip technique. The outer nucleus is easily removed—mostly with aspiration but with a little bit of emulsification (Figures 14–31, 14–32, and 14–33).

The Hinge Method

The principles of the physiologic trampoline also apply to the hinge method and the rotation method. The hinge method also shares some characteristics of the cracking techniques.

After central sculpting is completed, attention is devoted to the inferior portion of the inner nuclear girdle. It is emulsified little by little using the principles of Spring Surgery until the girdle's wall opens and there is communication between the sculpted area and the cleavage plane between the inner nucleus and the outer nucleus (Figures 14–34, 14–35).

The inner nucleus is then rotated 180 degrees and the same thing is done to the inner nuclear girdle inferiorly again, creating two breaks in the girdle wall 180 degrees apart. While doing this, the floor of the nucleus is shaved constantly, thinning out the floor and leaving two separate halves of nucleus separated only by a thin bridge of posterior plate (Figures 14–36 through 14–42).

At this point, if this were cracking techniques, the two halves would be separated until the floor of the nucleus cracks and separates. The cracking techniques unfortunately expose the posterior capsule relatively early in the operation, which is not entirely necessary for soft to medium nuclei. Instead of using two instruments to separate the two halves of the nucleus outward, putting pressures against the capsular bag 180 degrees from each other, we use the thinning of the posterior portion of the nucleus as a hinge and allow the two halves of the nucleus to come in toward the middle of the capsular bag.

Obviously, all of this is relative. If the posterior plate is relatively thin, very little outward movement is needed to crack the plates apart and very little tension is placed on the bag itself. But, in the hinge method of Spring Surgery, we will try not to crack at all.

After the inner nuclear girdle has been split at two points 180 degrees apart, the nucleus is rotated 90 degrees so that the phaco tip faces one half of the nucleus. The tip is impaled into the nucleus and attempts are made to aspirate it toward the middle of the capsular bag. If the posterior plate is thin enough, the nucleus will fold in away from the bag and come to its middle, where it is emulsified quite easily. After half of the nucleus is removed, the other half is rotated 180 degrees and removed in the same fashion (Figures 14–43 through 14–52).

It is sometimes necessary in medium–hardness cataracts to quarter the nucleus before pulling the halves in. However, they pull in spontaneously in this technique, bending the posterior plate rather than cracking the plates apart and separating the posterior plate.

The spring component of the operation is most prominent during the breaking of the inner nuclear girdle 180 degrees apart. During both of those steps, the physiologic trampoline monitors and controls how much tissue can be removed safely with a single pass.

Preservation of the posterior plate means that smooth tissue is up against the posterior capsule and that if the nucleus does something unexpected, the posterior capsule is protected from being torn. This adds a margin of safety not found in the cracking techniques because, after one cracks, the posterior capsule is exposed and, if a nuclear piece should tumble or chatter, it could break the posterior capsule.

The posterior plate in the hinge method of Sping Surgery usually follows removal of the second half of the nucleus. If floats up freely and is emulsified easily, leaving behind only the outer nucleus, which can be aspirated, massaged, teased, and removed from the eye exactly the same as in the rotation method.

Summary

Spring Surgery is a variant of Dr. Fine's Chip and Flip technique and is heavily influenced by his previous work. Like the Chip and Flip, the posterior capsule is covered during the entire duration of the operation by the outer nucleus, which is like having a thick layer of foam rubber separating the firm inner nucleus from the capsular bag.

The difference between Spring Surgery and the Chip and Flip is subtle, reflecting a lesser dependence on the use of instrumentation to move the nucleus and a greater reliance on the physiologic trampoline. The capsular bag and the zonules become an important and telling guide to the pressures exerted in the eye and become a helper in reducing all of the surgical forces, that could cause problems during phacoemulsification.

Section VII

ENDOCAPSULAR PHACOEMULSIFICATION USING THE CRACKING PROCEDURES

15

Fractional 2:4 Phaco

David M. Dillman, MD
William F. Maloney, MD

Introduction

We love Yogi Berra stories, and here's one of our favorites! Yogi goes into a pizzeria and orders a personal sized pizza. "Would you like that cut into four or six slices?" asked the waitress. "Oh, you better make that four," replied Yogi, "I don't think I could eat six!"

Well, it may not make much difference how you divide a pizza, but it makes a whole lot of difference how you divide a cataractous lens. We'd like to share some thoughts with you about that very topic.

A telephone book contains thousands and thousands of facts, but not a single idea! On the other hand, Howard V. Gimbel, M.D., of Calgary, Canada, is an ophthalmologist full of wonderful and often breakthrough ideas in the field of cataract surgery. Dr. Gimbel deserves full credit for the idea of purposely and controllably breaking a cataract into a varying number of pieces via his Divide and Conquer technique.

Early on, John Shepherd, M.D., (we apologize to any others we may have omitted), embraced the concept and made appropriate adaptions to personalize the technique.

We first became aware of Dr. Gimbel's Divide and Conquer in late 1986 and Dr. Shepherd's adaptions in early 1988. We began the transition from the pupillary plane Three Steps to Phaco approach to a dividing approach later in 1988. Together we have tried numerous variations on the theme and have settled upon what we like to call Fractional 2:4 Phaco. "Three Steps to Phaco" was originated and developed by William F. Maloney, M.D.

Goals

The goals of Fractional 2:4 Phaco are first to cleave the cataract into two distinct zones and then to break the inner zone into four distinct quadrants. Each isolated quadrant is then safely removed within a specific "phaco zone" we call the Circle of Safety.

Capsulorhexis

The new hero of cataract surgery is the anterior capsule, and the new darling of cataract surgery, with good reason, is the capsulorhexis anterior capsulotomy. Capsulorhexis is physiologically, anatomically, and esthetically appealing. Its impact upon cataract surgery in general and phacoemulsification in particular has been dramatic, to say the least.

Capsulorhexis is at the foundation of Fractional 2:4 Phaco, because it is a phaco technique that quite definitely is "three-handed," with a broad rim of intact, strong symmetrical anterior capsule serving as our third hand.

Also, it's comforting to know that a broad rim of intact anterior capsule with its zonular attachments will provide a very suitable "bail out" position for ciliary sulcus IOL implantation in the (highly unlikely) event of a severely disrupted posterior capsule.

There are multiple good and effective capsulorhexis techniques that have already been covered in previous chapters. However, capsulorhexis is new enough that still to be completely answered is the question of whether an optimal size exists for the capsulorhexis opening.

Notre Dame football coach Frank Leahy always felt that "prayers work better when the players are big." While that might be true for football, how about capsulorhexis? Can it be too big? Can it be too small? Indeed, numerous variables must be considered, such as the pre-existing condition of the capsule/zonules; the behavior of retained subcapsular epithelium and retained cortical fibers; and the size, shape, construction, and material of the implanted IOL.

As of this writing, our goal is to create a symmetrical, central capsulorhexis opening that will cover the outer anterior surface of the implanted IOL by about one-half millimeter all the way around. For instance, we attempt to make a 5.0–mm–diameter capsulorhexis for use with a 6.0–mm optic.

Creating Two Zones

Golda Meir reminded us, "Don't be humble, you're not that great." As we advance along the continuum of phacoemulsification techniques, we are often surprised by the new things we learn about the human lens. One of these has been the existence of a third element in addition to the nucleus and cortex. Sandwiched between these is a transitional zone, a transitional material, if you will, that we have come to call the cortical zone. Others call it "epinucleus," while others yet call it "outer nucleus." We are confident we are all talking about the same thing, which is a zone of material that has matured beyond the cortical fiber stage but not yet to the nuclear stage. Perhaps our friend Marc Michelson, M.D., has the best nomenclature for it. Marc calls it " 'tain't" material. He says that it 'tain't nucleus and it 'tain't cortex!

In order to isolate the cortical zone from the true central nucleus, which we call the nuclear zone, and the true cortex, two separate maneuvers are required. Some surgeons call our nuclear zone the "endonucleus" or the "inner nucleus."

First is hydrodissection. Most authors seem to agree upon the term hydrodissection, but not upon the definition of its goal. We have often heard that hydrodissection separates the capsule from the cortex; we disagree. If that were true, there would be no need to do I/A after the phaco. We believe the real goal of hydrodissection is to separate the true cortical fibers (cortex) from the cortical zone (and to be more specific from the outer limit of the cortical zone) (Figures 15–1 and 15–2).

Secondly, because we desire to demarcate the nuclear zone from the cortical zone, we have chosen to call this process hydrodemarcation.

Before this gets too confusing, perhaps now would be an opportune time to mention the processes of hydrodelineation and hydrodelamination. As we understand them, they are processes that use ultrasonic cannulas in an effort to actually separate different zones or lamellae within the nuclear zone itself. Hydrodemarcation simply isolates the nuclear zone as a single entity and separates it from the inner lining of the cortical zone. We love to call this particular cleavage plane the demarcation zone, or (are you ready for this?) the DMZ. Like all DMZ's, nothing aggressive will take place in this DMZ.

Following fully successful hydrodissection and hydrodemarcation, we are left with totally separated and free nuclear and cortical zones. But some days you feel like the pigeon, and some days you feel like the statue. Sometimes it's just not possible to obtain the cleavage planes desired or with as much totality as wanted. This is, of course, because each cataract has its own personality and composition. But in the vast majority of instances, proper technique and patience result in successful hydrodissection and hydrodemarcation.

Our technique for hydrodissection is much as has been previously described. We place a blunt 26–gauge cannula on a 3–cc BSS syringe just under the rim of the anterior capsule in the right inferior quadrant and slowly inject. This may be repeated in other quadrants until a good posterior fluid wave is visually identified. One of us (DMD) is also very fond of supplementing traditional hydrodissection with a special superior hydrodissection using a small, acutely curved Binkhorst cannula.

Hydrodemarcation is done with the same hydrodissection 26–gauge cannula. In general, the cannula is directed into the substance of the lens in the mid–periphery of the right inferior quadrant of the cataract. How deeply? As deeply as the cataract will allow. As a rule, we suggest you advance the cannula until you actually start to push the eye away. (With the very soft lenses, the cannula is inserted more into the center of the lens and a bit more gingerly to about one-half the thickness of the lens.)

Howard Fine, M.D., has taught us that it is first wise to slightly withdraw the cannula within the same track to allow easier injection. Once slightly withdrawn, BSS is slowly injected. Again, we are seeking visual evidence of a fluid cleavage plane between the nuclear and cortical zones that is often manifested by a delightful "golden halo" created by the red reflex around the nuclear zone (Figure 15–3). This process might also need to be repeated in other quadrants.

Creating Two Halves

At this point, we now go about the process of breaking the nuclear zone into four equally sized quadrants. Ideally, the isolated cortical zone would remain intact to be removed as a separate entity following removal of the nuclear zone. However, Mary Poppins rarely does cataract surgery, so the ideal doesn't always take place. Thus, we'll explain variations on the ideal as we go along.

The first maneuver in the creation of quadrants is fashioning a central valley to bisect the nuclear zone. The central valley is a furrow that should extend very deep into the nuclear zone and peripherally to the DMZ (Figure 15–4).

How deep is deep? For most lenses, at least 80% of the depth of the nuclear zone; for really firm, brunescent lenses, closer to 90% to 95%. Deep! But aren't you afraid of breaking the central posterior capsule, you might ask. Hopefully, not sounding too cavalier, the answer is no . . . because of the broken watch theory. Even a broken watch is correct twice a day! However, as phaco surgeons, we do not have that luxury, thus continuous attention to detail is our mandate. By paying close attention to the many depth indicators experienced phaco surgeons have available in their arsenals, the central posterior capsule should not be threatened (Figure 15–5).

Please keep in mind that the central valley is made in only nuclear zone. Below that lie the DMZ, the cortical zone, the cortex, and then the posterior capsule.

Peripherally is where more caution needs to be taken.

Apparently, there was at least some humor in the Kremlin before Gorbachev and Glasnost. Nikita Khrushchev was fond of telling the story of the Russian citizen who broke into the Kremlin shouting, "Khrushchev is a fool, Khrushchev is a fool!" He was arrested and placed in prison for 23 years—three years for illegally entering the Kremlin, Khrushchev would say, and twenty years for revealing a state secret.

Well, let us tell you a state secret about Fractional 2:4 Phaco. We have both learned that it is the inferior peripheral posterior capsule that is more threatened in Fractional 2:4 Phaco than the central posterior capsule. Why? Because it is an area in which most of us are not accustomed. The triangle of safety used in the Three Steps to Phaco absolutely forbids us from coming anywhere near the inferior peripheral capsule. But proper formation of the central valley will take us down there, so let's keep several things in mind.

First, remember the natural concavity of the underside of the cataract. As our phaco strokes go deep and wide, we must taper them in accordance with the concavity of the area. Simply plowing straight ahead, once fairly deep, is a path to damnation! Well, at least a path into the vitreous.

Secondly, our old friends the DMZ and the cortical zone. As we approach the peripheral extend of the nuclear zone (which we will do in low power, linear phaco), the DMZ will often open and reveal the cortical zone. We use the cortical zone in much the same fashion an outfielder uses the warning track in baseball. Stop before you hit the wall.

Once the first half of the central valley is complete, the second instrument (for us a cyclodialysis spatula) is introduced through the side port incision, and the nuclear zone is spun counter-clockwise 180 degrees (Figure 15–6). The ease with which this is done is dependent upon the success of the previous hydrodissection and hydrodemarcation. (In the same vein, again depending upon the hydrodissection/hydrodemarcation, the nuclear zone/cortical zone might spin as one unit, or the nuclear zone may spin independently of the cortical zone.) If the nuclear zone does not spin with only minimal effort, all instruments should be withdrawn and the hydrodissection/hydrodemarcation process repeated.

The second half of the central valley is done according to the same principles as the first half (Figure 15–7). We purposely chose the word "valley" to create a mental image of a furrow with deep, tapered walls. The top of the valley should only be about two phaco tips wide, and the walls should be tapered in such a fashion that the bottom of the valley is about one phaco tip wide. "Steep and Deep" is the central valley motto.

First Fracture

Neil Sedaka's popular song states that "breaking up is hard to do," but with a proper central valley, breaking the nuclear zone into two equal halves is amazingly easy if you follow a few simple rules (Figure 15–8). First, place the phaco tip (you're now in the foot position 1) and spatula into the bottom of the central valley—not the top, not the middle, the bottom. Then using a cross-action technique, physically split the nuclear zone, but do so respecting the natural concavity we spoke of earlier. The movement is not so much a lateral movement as it is a down and out movement (Figure 15–9).

"Are you badly injured?" the police officer asked the victim of a traffic accident. "How should I know?" she responded. "I'm a doctor, not a lawyer."

Aren't you afraid of injuring the posterior capsule by splitting the cataract in two? Again, no—not if the above-mentioned principles are strictly adhered to. By creating a deep and steep central valley, placing the instruments into the bottom of the valley and moving them in a down and out fashion (following the natural contour of the posterior capsule), very little stress is placed upon the capsule or the zonules. In fact, Miyake-type video observation reveals that this nuclear zone splitting is significantly less stressful to the capsular/zonular complex than implanting a 14.00–mm IOL into the capsular bag!

All that needs to be split is the nuclear zone. At times, the cortical zone will split in concert with the nuclear zone. If it does, fine. The remainder of the procedure is affected very little one way or the other. Whether it be whole or two halves, the cortical zone will continue to serve as a protective "cushion" for the posterior capsule.

Half 1, Quarters 1 and 2

The cyclodialysis spatula now spins the halved nuclear zone 90 degrees, bringing the first half (H1) into proper position for quartering (Figure 15–10).

The quartering process begins by creating yet another furrow, or secondary valley, in the very center of H1.

Former Attorney General Edwin Meese once recalled the advice of his predecessor, William French Smith, regarding tough days. The outgoing attorney general warned Meese that there would be many a day when he would feel like "the javelin competitor who won the toss of the coin and elected to receive."

Early on, we had some tough days creating this secondary valley. Then we hit upon a key. The first pass of phaco here should start, not at the top of H1 (as we traditionally always teach), but as close to the bottom of H1 as the firmness of the nuclear zone will allow.

Starting deep allows for two niceties. One is that H1 is easier to control, and two is that H1 will, at times, spontaneously split into quarters with the very first pass of this deep phaco.

However, in most instances, a second phaco pass is necessary just above the first pass to unroof it. The peripheral extend of the secondary valley need be only closely approaching the DMZ. If it should extend all the way to the DMZ, fine. We still have our cortical zone warning track to protect us.

Having created a deep and steep secondary valley, the spatula and phaco tip are placed into the furrow and, with minimal force, are used to split the first half (H1) into the first two quarters (Q1, Q2). Because of the smallness of the secondary valley, the movement of these two instruments now is straight outward, as opposed to the cross-action maneuver of the first split (Figure 15–11).

As mentioned in the introduction, we have gone through multiple variations with this type of phacoemulsification. One has to leave Q1 and Q2 alone at this point, spin the second half (H2) into position, and create the third and fourth quarters (Q3 and Q4). We ended up not liking that option because of a control quality issue (as opposed to a quality control issue). We found it better i.e., easier to maintain overall control, to remove Q1 and Q2 before approaching the second half. So, that's what we're recommending; trust us.

Bowling professional Don Carter feels that "one of the advantages of bowling over golf is that you very seldom lose a bowling ball." We feel that one of the advantages of Fractional 2:4 Phaco is that you very seldom lose a posterior capsule or a significant number of corneal endothelial cells. That's due in large part to the "Circle of Safety." The circle of safety is a central phaco zone that is pretty much defined by the capsulorhexis. The remainder of the phacoemulsification will be strictly confined to within the circle of safety and primarily within the inferior half of the circle of safety (Figure 15–12). This will place us far from the underside of the cornea and far from the posterior capsule, which is being further protected by our cortical zone safety net.

The first quarter is grasped with the phaco tip and brought into the circle of safety. Here it is emulsified. Now, just as the ukulele is the missing link between noise and music, low linear phacoemulsification and maximal utilization of the second instrument are the missing links to success with removal of each quarter (Figure 15–13).

With proper hydrodemarcation, we will most often remove only nuclear zone Q1. At times, a small peripheral part of cortical zone Q1 will spontaneously be aspirated away. But in the vast majority of cases, most of the central portion of cortical zone Q1 will remain intact.

The second quarter is then brought into proper position within the circle of safety by careful manipulation utilizing the spatula and phaco tip. Just as Q1, it is emulsified, attempting to preserve as much underlying cortical zone as possible (Figure 15–14). We're halfway home.

Half 2, Quarters 3 and 4

The spatula engages H2, places it securely into the superior capsular fornix and, respecting the natural capsular circumference, spins it 180 degrees counterclockwise so that it now occupies the inferior half of the capsule (Figure 15–15). Here we go again.

A young girl turns in her composition entitled "Our Cat." The teacher returns it the next day with a big, fat, red "F" scrawled across the top with the comment, "This is the same composition your sister turned in last year!" "Why not," replied the little girl, "it's the same cat!"

Well, the second half is a little bit of a different cat than the first half, primarily because the capsular bag is now half empty, giving more room for H2 to move about. This simply means we need to slow things down a bit (Remember the old saying that "people forget how fast you did a job, but they remember how well you did it."), maximize the abilities of both instruments, and stay within the circle of safety. Other than that, H2, Q3, and Q4 are dealt with in an identical fashion as was H1, Q1 and Q2.

Cortical Zone

The difference between ordinary and extraordinary is that little extra. It's all those little extra things we've done up to now that have helped us preserve as much cortical zone as possible. With the nuclear zone safely removed, however, the cortical zone can now be evacuated using the phaco handpiece much more as a highfalutin irrigation/aspiration instrument than a phaco instrument.

We deal with the cortical zone in much the same fashion that Howard Fine, M.D., has described in his Chip and Flip technique and Paul Koch, M.D., in his Spring Surgery. Using the second instrument to both manipulate the cortical zone and protect the posterior capsule, the cortical zone is aspirated away, perhaps with occasional short bursts of very low power, linear phaco, but once again confining ourselves to the circle of safety.

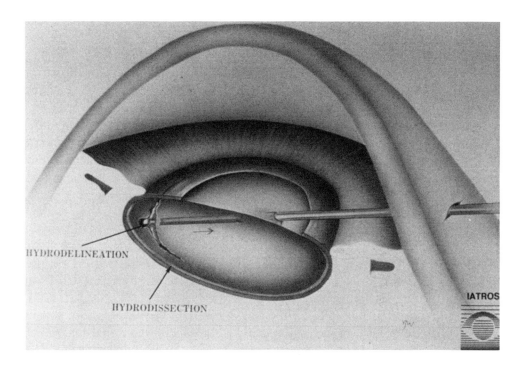

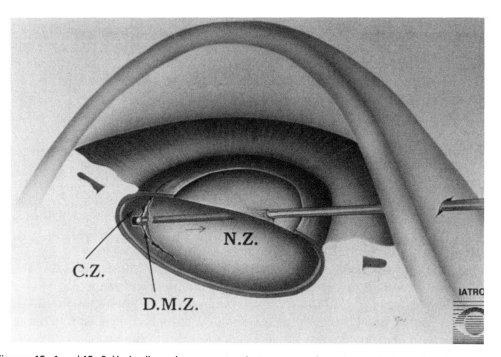

Figures 15–1 and 15–2. Hydrodissection separates the true cortex from the cortical zone (C.Z.). Hydro-demarcation separates the cortical zone from the nuclear zone (N.Z.). Both are accomplished by allowing BSS to find and break natural cleavage planes within the substance of the lens.

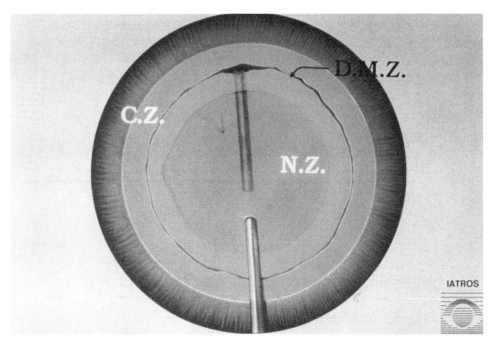

Figure 15–3. With successful hydrodemarcation, a clear space, the demarcation zone (D.M.Z.), becomes apparent between the cortical zone and the nuclear zone. Often the D.M.Z. manifests itself as a golden halo.

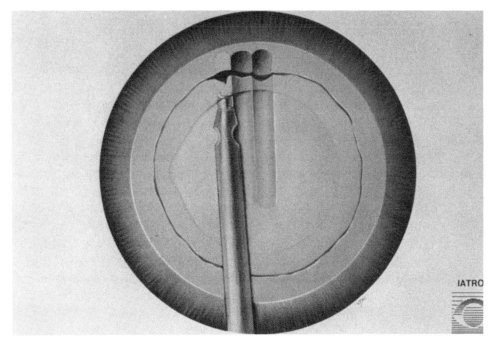

Figure 15–4. The peripheral extent of the central valley need only be to the D.M.Z., and in fact, we recommend just that. The beauty of isolating the cortical zone from the cortex and peripheral capsule is that it allows for a "warning track" to tell you to go no further. The above example illustrates carrying the peripheral central valley a little past the DMZ, but still well short of the peripheral capsule.

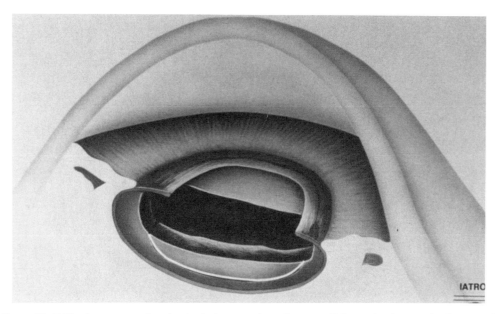

Figure 15–5. The key to stressless (on both the capsule and surgeon!) fracturing is to make the central valley as deep as safely possible. The more firm the nuclear zone, the deeper need be the central valley. While going deep, we use the underlying cortical zone as a "safety net" to keep the posterior capsule protected.

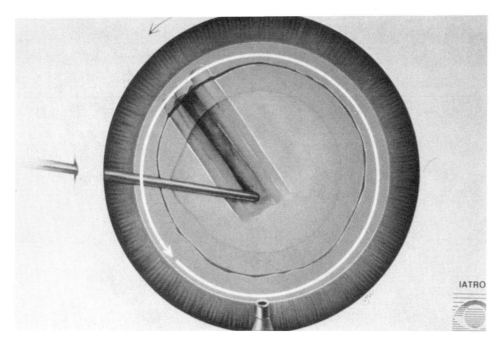

Figure 15–6. To complete the central valley, spin the nuclear zone 180 degrees in a counter-clockwise fashion. The ease with which this is done will be a direct result of the completeness of the hydrode-marcation. The extent to which the cortical zone accompanies the nuclear zone is tied more to the completeness of the hydrodissection.

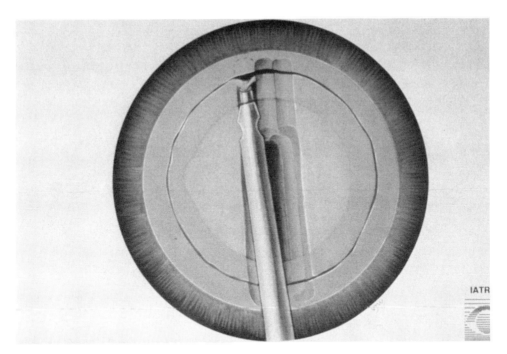

Figure 15-7. The second half of the central valley is completed by adhering to the same principles as the first half, remembering that "deep" is more important than "wide" . . . the peripheral capsule is actually more vulnerable than the posterior capsule.

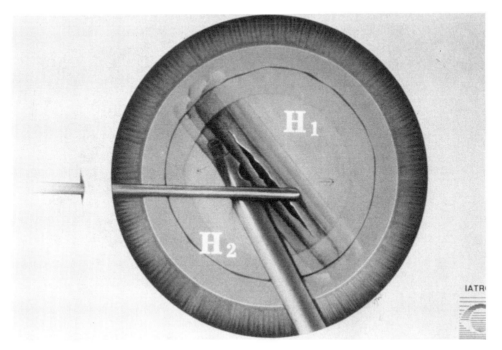

Figure 15-8. The first fracture breaks the nuclear zone into halves (H1 and H2). A deep and steep central valley is essential.

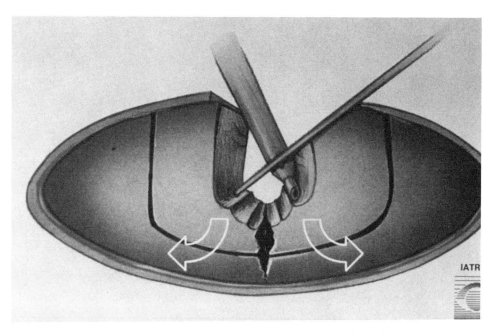

Figure 15–9. Stressless fracturing is facilitated by proper central valley formation, proper placement of the instruments, and respecting the natural concavity of the underside of the lens. The movement is "down and out" as opposed to purely lateral.

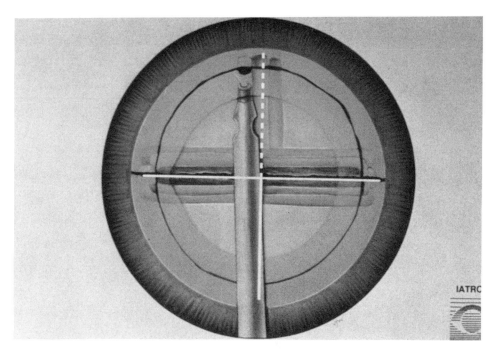

Figure 15–10. A secondary valley is made in the first half so that it may be further divided into quarters. Once again, "deep" is more important than "wide". The peripheral extent of the valley need only be approaching the D.M.Z., but if it does extend beyond the D.M.Z., as shown, use cortical zone warning track.

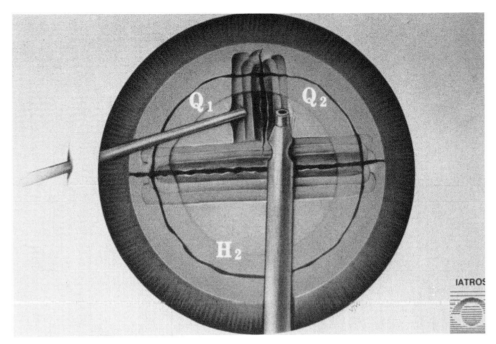

Figure 15–11. Half 1 is split into two quarters (Q1 and Q2). This is accomplished by placing the instruments into the bottom of the groove and moving them away from each other, as opposed to the cross-action movement of the first split. (Compare to Figure 8).

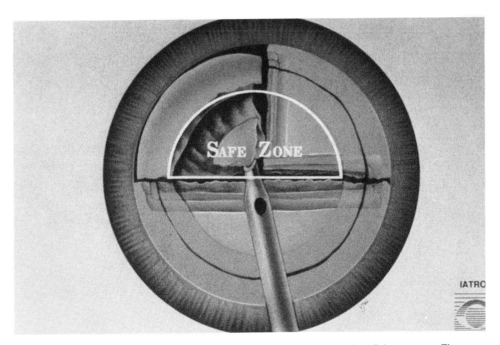

Figure 15–12. The circle of safety is the "phaco zone" for the remainder of the surgery. The surgeon should never be in the phaco mode (foot position 3) outside the circle of safety at this point. It's quite acceptable and, in fact, often necessary, to move the phaco tip outside the circle of safety when used as a maneuvering instrument, but when used as a phaco instrument, confine it to the circle.

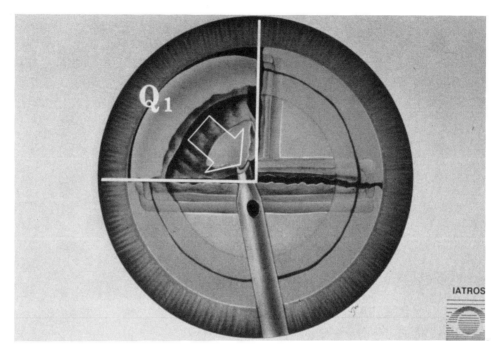

Figure 15–13. The first quarter is removed within the circle of safety, utilizing low linear phaco and pulsing (either manually controlled, by the surgeon's foot, or automated, by machine control). The second instrument, not shown here, is also key to controlled removal of each quadrant.

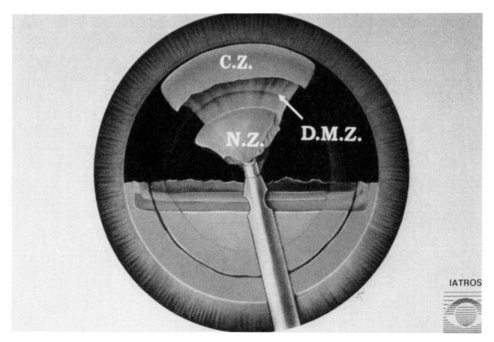

Figure 15–14. Whenever possible, effort should be made to remove only the nuclear zone of each quadrant, preserving as much cortical zone as possible to act as a protective cushion for the posterior capsule.

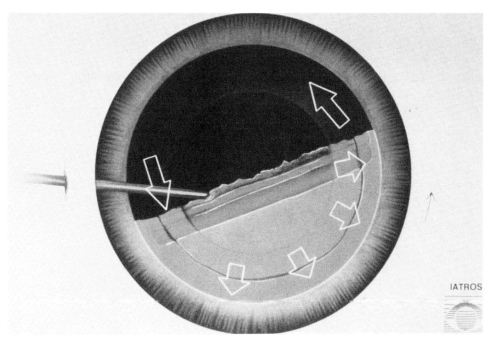

Figure 15–15. The second half is spun 180 degrees to bring it into proper position for repeating the process. Keeping this half "up against the wall", i.e., snug in the capsular fornix, facilitates its spinning.

Summary

As we all advance along this continuum we call phacoemulsification, it is wise to take the pragmatic approach. By pragmatic, we mean the perspective taken by tennis pro Ilie Nastase when he was questioned by a credit card company about his failure to report the loss of his wife's credit card even though it had been missing for more than a year and, in fact, turned out to have been stolen. "Whoever had it," explained Nastase, "was spending less than she was." Now that's pragmatism.

Advancing along the continuum necessitates identifying methodologies that are not only practical in their approach, but also offer a truly greater likelihood of preserving as much anatomy and physiology of the human eye as possible. We believe Fractional 2:4 Phaco, a variation of Dr. Gimbel's and Dr. Shepherd's technique, does just that. Separating a nuclear zone from a cortical zone and then using the cortical zone as a peripheral warning track and posterior safety net affords tremendous protection to the peripheral and posterior capsule. The use of a small, central capsulorhexis and circle of safety both gives added protection to the corneal endothelium and provides a wonderful environment for IOL implantation. Fractioning the nuclear zone into quarters allows for exquisite control over the phacoemulsification process and greatly facilitates emulsifying the "rock hard" cataracts and working through small pupils.

Asked if he is actually responsible for all the quotes attributed to him, Yogi Berra explained, "I really didn't say everything I said." Thanks, Yogi! 'Nuff said.

16

Posterior Nuclear Fracture and Quartering Technique

James A. Davison, MD

This method works well in any nucleus that is firm enough so that the cyclodialysis spatula can manipulate the central nucleus without sinking through too much. It works well on medium–firm lenses. The posterior peripheral nuclear layer of firmer 3 + lenses may be so hard that it will not fracture when the surgeon attempts to draw in the peripheral segment when using either of the first two strategies presented in Chapter 12 (Figure 16–1, 16–2).

This failure occurs for two reasons. First, the posterior layer is too rigid and too broad to fracture (Figures 16–3, 16–4). Second, good suction to fold the peripheral nuclear portion over is harder to obtain in firmer lenses. Small pieces of peripheral anterior material may be aspirated, but a large single piece of relatively unmanageable posterior disc and peripheral posterior nuclear material may be left. In these cases, fracturing the posterior nuclear plate into four pieces with a bimanual spreading motion, thus quartering the nucleus, makes each quarter nuclear fragment manageable. As with the removal of soft and medium–firm nuclei, there exist many variations of nuclear plate segmentation methods.

Howard Gimbel (Divide and Conquer) and John Shepherd (In Situ Fracture) were the first to promote the concepts of bimanual tearing of groups of nuclear fibers in a spreading–type fracture of nuclear layers and separation of consequent nuclear segments for more convenient emulsification. Dave Dillman and others have added their own personal touches to the basic technique. Gimbel, Shepherd, and Dillman all prefer the 30–degree tip for this procedure; I still like the 45. I think that the same principles of isolated cutting and suction and nuclear positioning and manipulation apply to these fracturing-quartering methods of removal as well as for softer lenses.

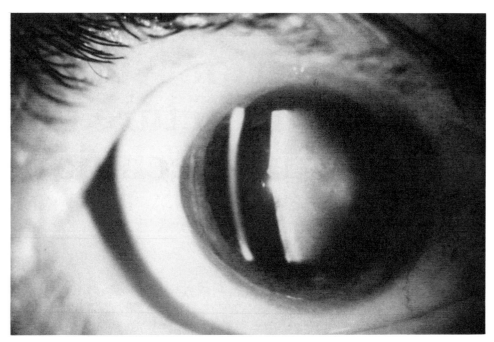

Figure 16–1. This 3+ firm cataract will be removed more easily with a posterior nuclear fracture-quartering technique rather than the rim reduction methods used for softer cataracts.

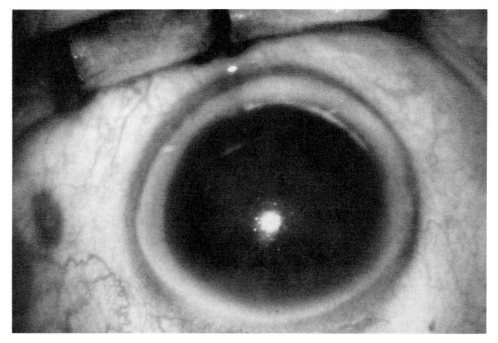

Figure 16–2. At the beginning of surgery, some cataracts look softer than stated in the clinical examination, which might cause the surgeon to consider an alternative operative plan. Actual firmness will be revealed during phacoemulsification.

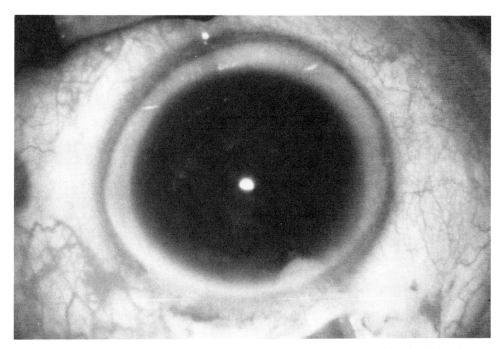

Figure 16–3. The mid–level and deeper nuclear layers are homogeneous and hard and tough to get through.

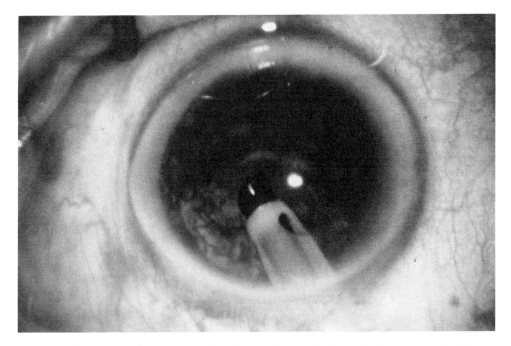

Figure 16–4. Even when through, posterior thinning does not help much. The surrounding firmness and strength of the peripheral nucleus will not permit a good suction hold to be applied to the rim. Infolding and fracture are very difficult unless a radial defect is made in the posterior nuclear plate to assist.

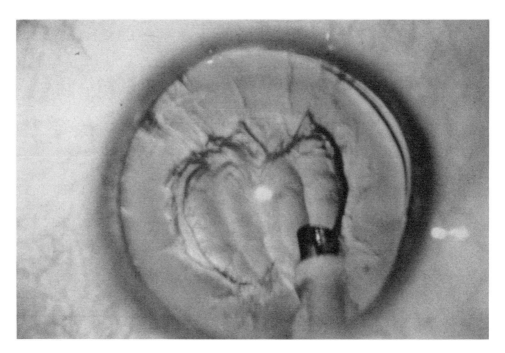

Figure 16–5. Nuclear sculpting is initiated but stopped early, with generous right and left sides remaining.

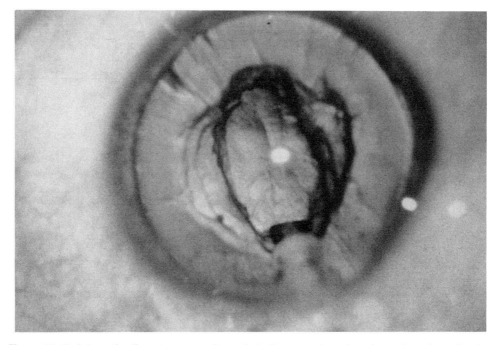

Figure 16–6. A two–tip diameter groove is made in the central nucleus. It needs to be quite deep, leaving just a thin portion of the posterior disc, which will be fractured and separated. I prefer the 45–degree tip for this because it allows improved visualization of the tip's apex. It is usually easier to accomplish the first rotation before the first fracture is made.

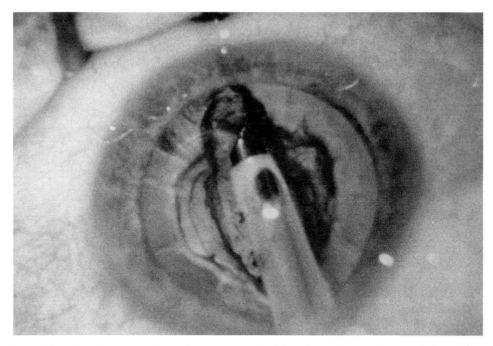

Figure 16-7. The 45-degree tip can be placed on its side and used in a cutting mode to get through most of the peripheral anterior rim. The surgeon does not usually need to go so far out, as in this case.

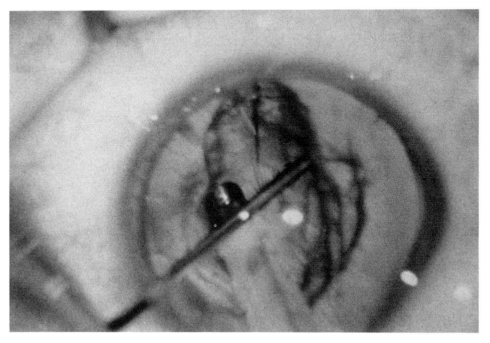

Figure 16-8. With the infusion on, a gentle spreading motion is used to fracture the posterior nuclear disc and separate the right and left inferior quarters.

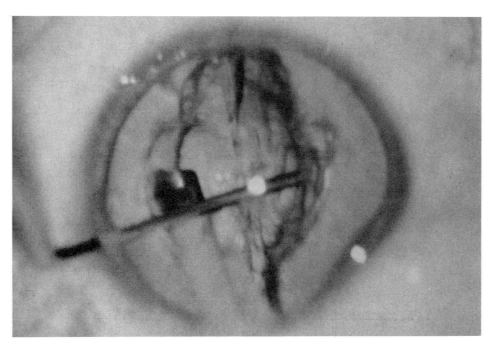

Figure 16–9. At times, the superior right and left halves will separate as well. However, the superior nucleus is usually rotated to the inferior position before being acted upon.

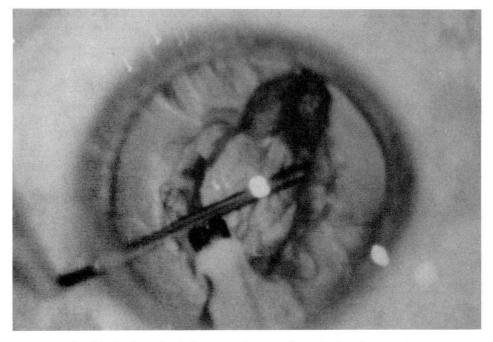

Figure 16–10. Good hydrodissection is important because the entire lens is rotated one quarter turn before significant peripheral rim debulking can occur.

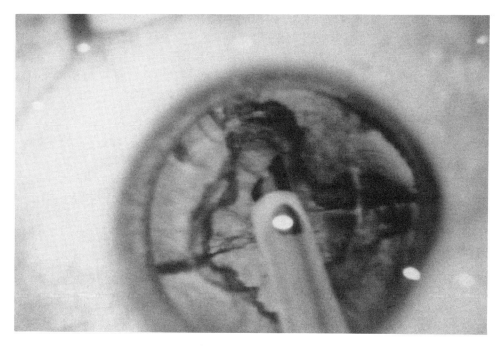

Figure 16–11. The left half of the nucleus is now in an inferior position where another deep groove can be created.

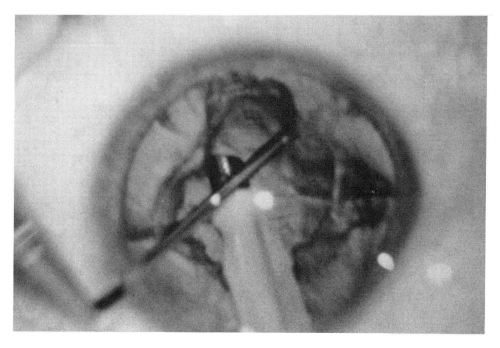

Figure 16–12. The phacoemulsification tip and cyclodialysis spatula are in position to spread the inferior right and left quarters apart.

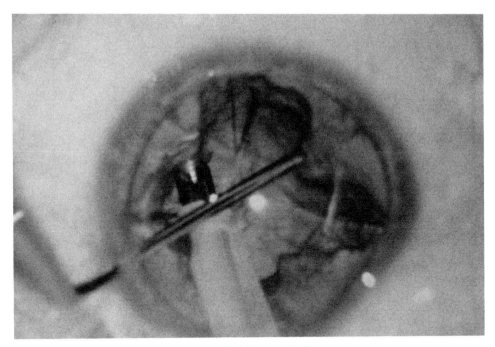

Figure 16-13. The right and left quarters are slightly separated.

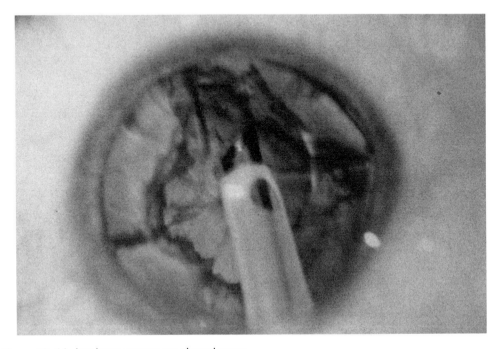

Figure 16-14. Any large corners are shaved away.

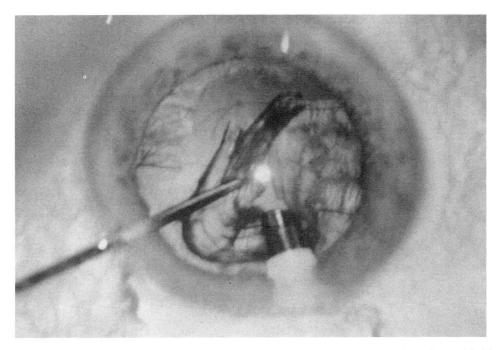

Figure 16–15. Large three–dimensional chunks of nucleus are undesirable and can be avoided by shaving down the central nuclear apices of the divided quarters.

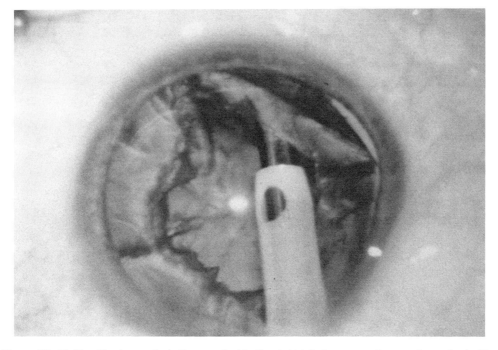

Figure 16–16. The 45–degree tip is then applied to the right inferior quarter in a suction attitude. The firm quarter nuclear rim can now be drawn centrally.

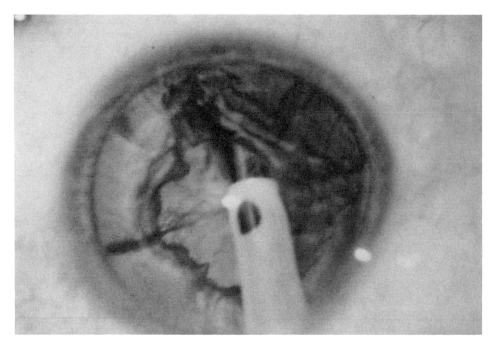

Figure 16–17. In this central position, the nuclear substance can be more safely aspirated with the assistance of brief taps of emulsification energy.

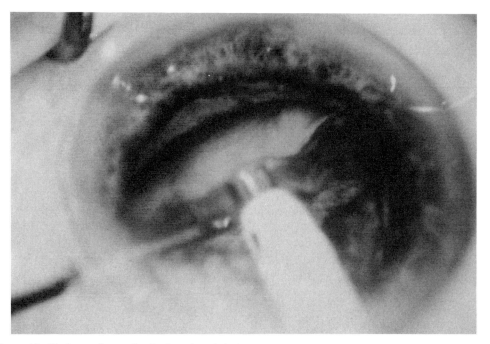

Figure 16–18. A very large chunk of nucleus is being rolled over by aspiration with a 30–degree tip. The large size of this nuclear fragment threatens the anterior capsule and corneal endothelium. (Photograph courtesy of David Dillman, M.D.).

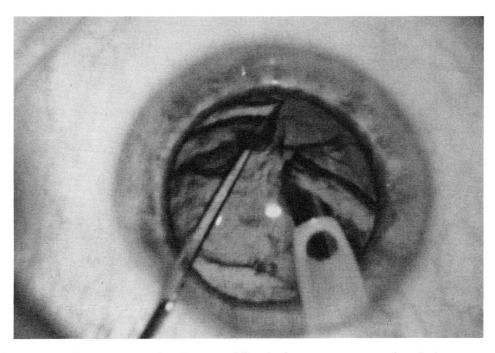

Figure 16-19. Extra care must be taken to stabilize the last two quarters so they don't escape or damage the capsular bag.

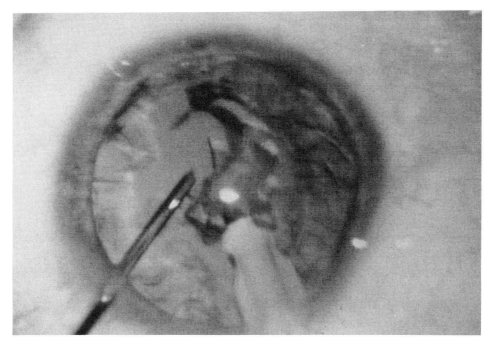

Figure 16-20. The small nuclear fragments are supported by the cyclodialysis spatula so that the phacoemulsification tip will not have to be too close to the posterior capsule while they are aspirated.

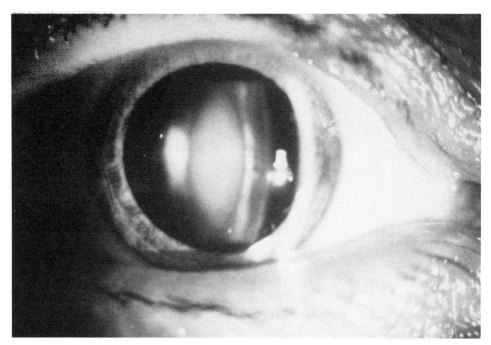

Figure 16–21. A 2+ nuclear cataract is ideal for the beginning phacoemulsification surgeon to do either a cutting-suction method or the CCS hybrid technique.

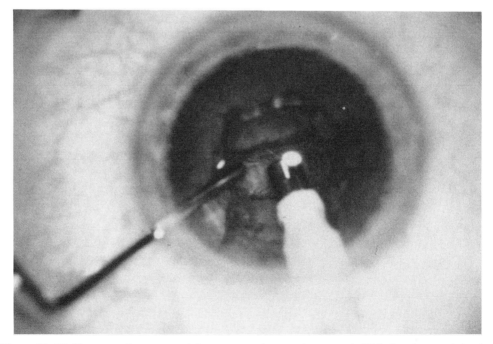

Figure 16–22. The second groove and fracture are about to be created. With firm lenses, it is also helpful to partially groove down the superior section so that the phacoemulsification tip does not push posteriorly on it while trying to gain access to the inferiorly positioned one.

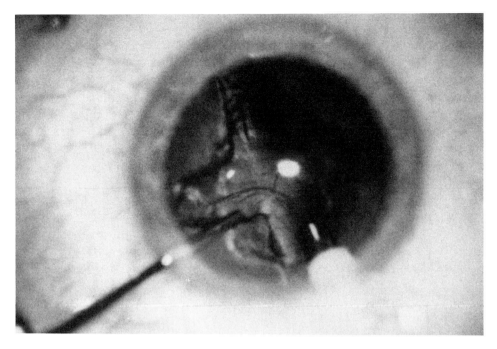

Figure 16–23. All four grooves and fractures have been created, and some deeper peripheral nuclear thinning has been accomplished in the inferior right quadrant.

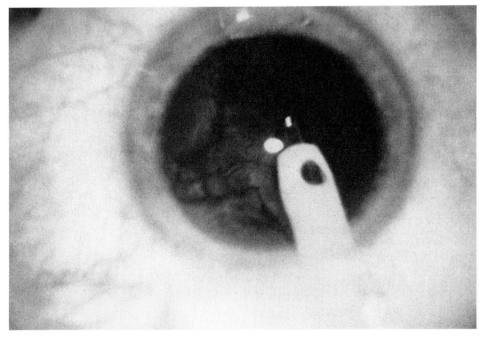

Figure 16–24. More deeper peripheral thinning is accomplished.

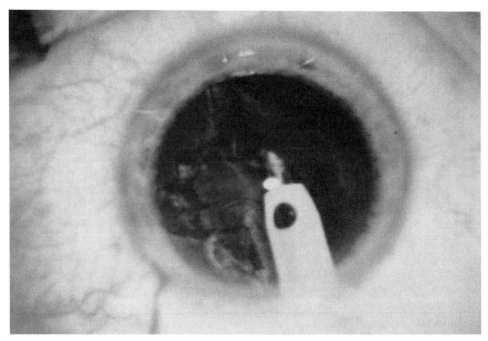

Figure 16-25. The thinner debulked nuclear fragment is drawn centrally in a forward roll by vacuum and brief taps of emulsification energy.

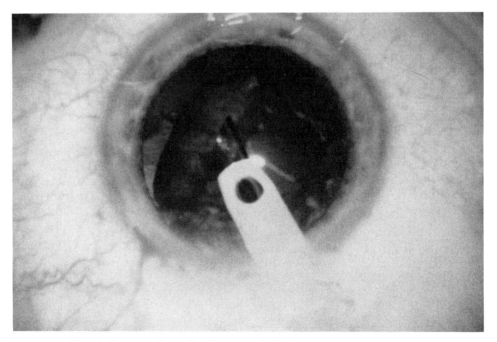

Figure 16-26. The relative two-dimensional nature of this firm nuclear fragment can be seen when compared to the larger (but softer) nuclear chunk in Figure 16-15.

Nuclear sculpting is accomplished, but a fair amount of material remains on either side of a groove cut down the middle of the nucleus (Figures 16–5, 16–6). An isolated cutting mode is used to thin the nuclear rim without aspiration (Figure 16–7). A gentle spreading motion is applied to each side of the groove, creating a small fissure between the two inferior quarters (Figure 16–8). This motion can be extended into the superior quarters as well, but it is only necessary to reach the center of the nuclear plate. I reserve this spreading motion for the groove that is rotated to the inferior position just after the groove has been extended into the inferior nuclear rim with the tip on its side in a cutting mode (Figures 16–9 through 16–13).

After the lens has been quartered by creating four small fissures (the first rotation of the nucleus), the remaining central sculpting is completed by shaving down some of the remaining central prominences (Figure 16–14). Many times, each segment is rotated into an inferior position where this cutting mode can be applied (the second rotation). Because this debulking is much harder to accomplish if any of the quarters has been removed first, it should be done before actual segment removal.

If these quarters are not debulked, large three–dimensional chunks of nucleus will remain for the suction mode (Figure 16–15). These large nuclear segments are more difficult to handle than the relatively two–dimensional curved tubular sheets that can be emulsified with the tip in the suction orientation after central thinning (Figures 16–16, 16–17).

Aspirating four quarters may require one last rotational component of the nucleus, the rotation of the superior nuclear fragments to the inferior position after the right and left segments have been removed in situ. As they are emulsified in the suction mode, they roll anterior and inward, and their sheer bulk or their rotating apices may threaten the posterior capsule, anterior capsular rim, and corneal endothelium (Figure 16–18). Separation and emulsification of the latter two quarters require stabilization with the cyclodialysis spatula so that the segments don't tumble or damage surrounding structures (Figures 16–19, 16–20). Appropriate positioning must be obtained before even low power emulsification energy is delivered. Part of the positioning is provided by the suction on the nuclear fragment provided by the vacuum built up through aperture tip occlusion.

Very Firm Lenses

Nuclei of 2 + firmness and greater can be emulsified in this efficient fashion (Figure 16–21). A 2 + firm lens, of course, can be removed with suction and cutting power without posterior nuclear cracking. Very firm lenses benefit from a slightly more rigid routine. All four central grooves and posterior nuclear fractures are created first in an initial rotation of the nucleus (Figures 16–22, 16–23). Fairly extensive deep posterior peripheral nuclear thinning is accomplished with the tip in its cutting mode during another rotation or part rotation (Figure 16–24). This extensive thinning facilitates the safe gentle anterior central rolling inward motion that is seen when brief taps of low power phacoemulsification energy combine with higher vacuum in the suction mode (Figure 16–25). The fragments are less chunky and more two–dimensional and, I think, are safer and easier to remove (Figure 16–26)

Section VIII

INTERCAPSULAR PHACOEMULSIFICATION

17

Endocapsular Phacoemulsification With Mini-Capsulorhexis

Marc A. Michelson, MD

Introduction

Endocapsular phacoemulsification is defined in this chapter as phacoemulsification of the lens nucleus through a small oval capsulotomy made at the superior aspect of the anterior lens surface. This technique facilitates near total compartmentalization of the emulsification process within the lens capsule, thereby providing significant protection to the iris and corneal endothelium.[1,2,3] Using this method, I have observed no endothelial cell loss (0% loss) in between 40% to 65% of my cases and minimal cell loss (4% or less) in 20% to 30% of the remaining cases. Following phacoemulsification, the anterior capsule is removed by capsulorhexis, resulting in a larger, continuous round capsulotomy prior to IOL insertion in the capsular bag.

This chapter will describe the technique of endocapsular phacoemulsification with emphasis on the evolution and application of the "mini-capsulorhexis," the technique of hydrodissection, nuclear emulsification/sculpting, and the prevention/management of potential difficulties encountered by the surgeon first learning endocapsular phacoemulsification.

Summary

Phacoemulsification is steadily becoming more and more popular. With its increase in prominence, ever-enlarging numbers of phaco surgeons worldwide search eagerly for advances in techniques to push the ideals of safety and favorable postoperative clinical results steadily closer to perfection. In short, how can phacoemulsification be brought to the pinnacle of efficacy? Enter endocapsular phacoemulsification, defined as emulsification of the nucleus within the confines of the lens capsule. When properly understood and utilized, this technique offers extremely promising benefits that can be readily achieved by today's phaco surgeons.

With endocapsular phacoemulsification, emulsification of the nucleus takes place through what I define as a "mini-capsulorhexis" opening near the superior iris border of the anterior capsule. This approach leaves the remainder of the anterior capsule completely intact during emulsification, facilitating complete and safe removal of the nucleus and cortex within the confines of the lens bag.

Endocapsular phacoemulsification through a mini-capsulorhexis offers many advantages:

1. It leaves intact a broad area of anterior capsule, which allows for "compartmentalization" of the phacoemulsification, thus creating a "physical barrier" that prevents nuclear fragments from escaping into the anterior chamber.
2. The intact anterior capsule protects the iris and corneal endothelium from turbulent forces created by the phacoemulsification.
3. The exchange of fluids between the lens capsule and the anterior chamber are kept to an absolute minimum, which additionally serves to maximize protection afforded by viscoelastics.
4. BSS, which enters the lens from the phaco tip, exits the eye in a flux surrounding the needle's shaft rather than creating turbulent currents in the anterior chamber. In fact, very little fluid enters the chamber unless the tip should approach the capsulotomy. This is demonstrated clinically by the stability of air bubbles under the cornea; a clear differentiation can be visualized under and over the anterior capsule.
5. An intact anterior capsule stabilizes both the nucleus and its fragments during phacoemulsification, thus preventing their entrance into the anterior chamber.
6. The overall risk of endothelial cell damage is significantly reduced.
7. This is a one-handed technique. It is unnecessary to introduce a second instrument from a side port incision, as there is only a relatively small opening in the lens bag.

Over the years, perhaps the greatest objection surgeons have had to adopting phacoemulsification has been their concern regarding trauma to the corneal endothelium. It is my opinion that this concern—extremely valid in the early days of phaco—has slowly and rightfully been addressed over the years, notably with the use of viscoelastics, as well as the evolution of phaco techniques that have shifted the point of emulsification from the anterior chamber to the iris plane to the posterior chamber. Now, I believe that endocapsular phaco-emulsification offers the next and perhaps ultimate step toward endothelial cell safety in phacoemulsification.

The Capsulotomy

The primary goal of any surgeon learning endocapsular phaco is mastery of the capsulotomy, the most critical step in the technique.

It is essential that the surgeon perform the capsulotomy in such a way that a broad area of central anterior capsule remain in place during emulsification in order to use the capsule as a physical barrier to protect the corneal endothelium and compartmentalize the turbulence. At the same time, the opening into the lens, which is positioned just below the superior iris border, should also allow easy access of the phaco needle into the lens bag as it passes through the limbal wound. This access should provide the room to maneuver comfortably 360 degrees during lens emulsification and cortical removal.

When I refer to "mastery of the capsulotomy" as the primary key to successful endocapsular phacoemulsification, I refer specifically to what I have termed the "mini-capsulorhexis," which I have found to be the best capsulotomy for the endocapsular phaco technique. This is a small oval capsulotomy with a smooth circumferential margin. As previously noted, the mini-capsulorhexis should be placed near to and below the superior iris border to allow the most stable capsular opening.

Generally, between a 4 mm—5 mm x 0.5 mm—1 mm capsulorhexis is best (Figures 17–1 through 17–4). The creation of a continuous tear margin, furthermore, minimizes the incidence of anterior capsule ruptures.

While it is important that the size of the capsulotomy be relatively small—approximately twice the diameter of the phaco needle—it is equally important that it not be too small. The aperture of the capsulotomy should be sufficient to allow the silicone sleeve of the phaco tip to enter the lens and be manipulated inside the capsule without placing stress on the anterior capsular margins.

Additionally, in order to prevent tears, the surgeon must be able to maneuver the handpiece routinely without risk of stressing the margins. It is essential to realize that the success of the endocapsular phaco technique depends upon preventing the anterior capsule from splitting into two large flaps (resultant from stress caused by the phaco tip's pressure against the edge of the capsulotomy). In addition to negatively affecting the efficacy of the endocapsular phaco technique, large flaps will often impede removal of cortex and may have to be removed prior to cortical clean-up. For these reasons and others that I will expand on later in this chapter, it is better that the capsulotomy be made a little too large rather than too small, especially in the early stages of learning this technique.

Before beginning the detailed description of the mini-capsulorhexis, I believe it will be useful for the aspiring endocapsular phaco surgeon to have a brief background in the evolution of this approach. Understanding the drawbacks encountered with other capsulotomy styles used in conjunction with the endo-capsular phaco technique—particularly the can–opener and continuous tear "envelope" capsulotomy, though to a lesser degree—will enhance the surgeon's insight into the proper application and benefits of the mini-capsulor-hexis.

Limitations of the Can–opener Capsulotomy in Endocapsular Phacoemulsification

The "can–opener" capsulotomy is, of course, the most basic of all capsulot-omy techniques. Multiple punctures made with a bent needle or cystotome in a "frown configuration" limited to the superior border of the lens may be used in the endocapsular technique, but pose major limitations of which the surgeon should be aware (Figure 17–5).

Primarily, use of the can–opener capsulotomy results in a high probability of radial tear, initiating from a point on the inferior capsulotomy margin and extending to the zonules in each direction. This type of rent, likely to occur in 30% to 50% of cases, will result in the splitting of the anterior capsule into two large flaps (Figure 17–6). For this reason, the can–opener style of enve-lope capsulotomy should not be used when performing endocapsular pha-coemulsification, as this complication greatly diminishes the in-the-bag ben-efits.

Several points should be made in this regard:

1. Even if broad capsular flaps do develop during the procedure, a small, although very limited, degree of protective barrier is still maintained, serving to contain a small percentage of the nuclear particles within the lens bag.
2. Capsular flaps will result in a free exchange of fluid between the capsular bag and the anterior chamber, eliminating the protective value of the anterior capsule as a physical barrier.
3. While emulsification of the nucleus may not be impeded, cortical removal in the presence of large anterior capsular flaps is much more difficult.
4. The curvalinear opening of a can–opener capsulotomy is made up of a series of contiguous small punctures. Sufficient stress placed by the phaco needle against any of these inferior margin puncture sites, which are, in essence, mini-tears, will quite easily result in the puncture extending to the equatorial zonules.
5. If a tear occurs from the lateral margin, it will extend superior-laterally toward the equator. However, superior-lateral rents are not usually of clinical significance and, thus, rarely impede the procedure.

It should be noted that lateral or inferior extensions of capsular rents gen-erally pose little risk of progressing into the posterior capsule. These rents are regularly encountered by extracapsular surgeons when a nucleus is expressed through a can–opener capsulotomy, as the capsular margins simply do not have the rigidity to withstand the forces of nuclear delivery without tearing to the equatorial zonules (Figure 17–7).

Often, following nuclear expression through a can–opener capsulotomy, the capsular margins can be observed to disappear under the iris border only to be limited by the equatorial zonules. Based on my observation of videotapes of cadaver eye surgery using the video techniques of Kensaku Miyake, M.D., I have noted that equatorial tears do not extend beyond the zonules during endocapsular phacoemulsification. Similarly, a lateral or inferior rent that occurs during endocapsular phaco does not automatically mean that a serious complication, such as a rent extending into the posterior capsule, will ensue. This was demonstrated by David J. Apple, M.D., who has noted that a lateral rent from the capsulotomy margin will not extend into the posterior capsule unless zonular dehiscence has occurred.[4]

Continuous Tear "Envelope" Capsulotomy Reduces Risk of Inferior Capsular Rents

In light of all the difficulties encountered with the can–opener approach in endocapsular phaco, it rapidly became apparent to me that a continuous tear capsulotomy would provide the greatest degree of resistance to an inferior capsular rent. As a result, I very naturally evolved two styles of continuous tear capsulotomy during my development of this technique: the "envelope" style and the "mini-capsulorhexis" style. Both offer advantages, but, as I will explain, the mini-capsulorhexis presents the surgeon with the greatest advantages and the fewest drawbacks.

The envelope capsulotomy is initiated with a small puncture made with the cystotome in the capsule at one end, the instrument then swept laterally, arcing in a slight "frown" configuration, to create a smooth continuous tear (Figures 17–8 through 17–10). If the inferior edge of the continuous tear capsulotomy remains intact throughout the phacoemulsification, only the lateral margins can extend into the equatorial zonules. The "frown" orientation ensures that these extensions are directed superiorly.

When emulsification has been completed using an envelope capsulotomy technique, an intraocular lens may be inserted through this linear capsulotomy without risk to the zonules virtually all of the time.

Following IOL insertion, the central anterior capsule is removed. Next, a small incision is made with intraocular scissors in the anterior capsule. This becomes the starting point for the capsulorhexis, resulting in a three-sided continuous tear capsulotomy with only lateral and inferior margins (Figure 17–11).

The disadvantage to this approach lies in the fact that an intraocular lens that is bound by a three-sided, discontinuous capsulotomy may decenter superiorly. While this is not, in most cases, clinically significant, it does pose a more significant risk of stimulating decentration when a lens requiring capsular bag placement is used. This "three-sided" capsulotomy, shaped like a horseshoe, is subject to the concentric contraction of the capsule, resulting in the tendency for upward decentration of the lens at the superior margin.

One other point: In the occasional case in which a patient has a small pupil or a soft, milky cataract, this horseshoe-shaped envelope capsulotomy offers the surgeon a benefit. Specifically, if it is impossible to see the broad area of the anterior capsule and the surgeon does not want to perform a sector iridectomy to split the iris, the envelope capsulotomy works well without disturbing the iris sphincter. With patients who have soft, milky lenses in which a red reflex cannot be seen, it is often very difficult to see the capsule. Therefore, it may be prudent to sacrifice the ability to complete a 360–degree capsulorhexis, performing instead the endocapsular phaco technique under the anterior capsular flap through an envelope–style capsulotomy.

Mini-Capsulorhexis Offers Most Stable Capsulotomy for Endocapsular Phacoemulsification

The mini-capsulorhexis is the capsulotomy that preserves the integrity of the anterior capsule as much as possible. Because the "mini" continuous tear capsulorhexis is placed near the superior iris border, an extremely stable capsular opening for endocapsular phaco is created, providing the surgeon with the advantage of excellent tensile strength of the capsular margin. This modified capsulorhexis works very well and is, in my opinion, the ideal capsulotomy technique for endocapsular phacoemulsification, serving to minimize the incidence of anterior capsular rupture.

To reinforce what was stated earlier in this chapter, the mini-capsulorhexis should be large enough to allow comfortable entry of the phaco tip into the lens bag; generally, a 4 mm—5 mm x 0.5 mm—1 mm oval is ideal.

Starting the Mini-Capsulorhexis

A bent needle is introduced into the anterior chamber, which was previously filled with viscoelastic. The point of the initial puncture should be slightly inferior to the center of the planned capsulotomy. The needle is then swept laterally about 3 mm—4 mm. An upward push with the needle then creates a small vertical flap that should not exceed 0.5 mm in length (Figures 17–1 through 17–4).

Completing the Mini-Capsulorhexis

Completing the capsulorhexis with the bent needle is sometimes difficult since the capsulotomy is small. The surgeon must be aware of the possibility that the superior border may tear uncontrollably into the 12 o'clock zonules as the final arc completing the capsulorhexis is attempted. However, as with the envelope capsulotomy, a tear that extends under the superior iris border will very often not interfere with the surgeon's ability to perform endocapsular phaco, since a broad anterior capsular surface still remains intact.

By using capsule forceps, the capsulotomy can be completed with more precise control. First, the scleral-limbal wound should be enlarged with a pre-calibrated phaco keratome. Viscoelastic should be used as needed to maintain anterior chamber depth. Capsule forceps are introduced through the enlarged wound to grasp the free edge of the capsular flap, which is then torn horizontally to create the inferior margin of the capsulotomy. The flap is extended superiorly for 0.5 mm and immediately curved in the opposite horizontal direction to create the superior border. Care must be taken to ensure that the original puncture site is not left on the margin, but rather incorporated in the capsulorhexis that is removed (Figure 17–3). Otherwise, a microtear will exist on the edge of the capsulotomy, thereby precluding the accomplishment of a continuous tear capsulorhexis. This will leave open the risk of the tear extending radially into the zonules.

As excellent a technique as the mini-capsulorhexis is, the surgeon must bear in mind that it is not a complete and total fail-safe against capsular tears. Even the mini-capsulorhexis capsulotomy is not immune from splitting when exposed to the undue stress from forces of the phaco tip.

In the event the surgeon should visualize even the smallest rent developing in the inferior capsulotomy margin (Figure 17–12), the phaco tip should immediately be withdrawn from the eye. Because tears often start out extremely small, the secret is to stop performing phaco as soon as a rent is recognized so that corrective action can be taken while options remain open.

After the phaco handpiece has been removed from the eye, the I/A tip should be assembled and reinserted into the bag with the port facing up. The edge of the capsular flap is aspirated and occluded into the port. The I/A tip should be used to tear the flap back into the border of the original capsulotomy margin, thereby recreating a smooth, continuous capsulotomy (Figure 17–13). This technique will effectively obviate the likelihood of the rent extending, as the tear is completely eliminated. Depending upon the course and size of the rent, the opening of the capsulotomy will be somewhat larger following this procedure. However, even in a "worst case scenario," the surgeon should be aware that he or she will simply be in the same situation as if a larger, standard central capsulorhexis capsulotomy had been performed originally (Figure 17–14).

Another approach the surgeon might initially consider in the management of a tear in the capsulorhexis is to instill a viscoelastic into the anterior chamber using small capsule forceps to correct the rent. However, this technique runs the risk of extending the tear to the zonules, resultant from the combined forces of the insertion of the viscoelastic and the forceps placing increased pressure on the tear. Hence, it is not recommended.

When the phacoemulsification proceeds smoothly and is completed with the mini-capsulorhexis/small continuous capsular preserved, full protection to the corneal endothelium is achieved by the barrier effect of the anterior capsule. Prior to IOL insertion, however, the central anterior capsule should be removed

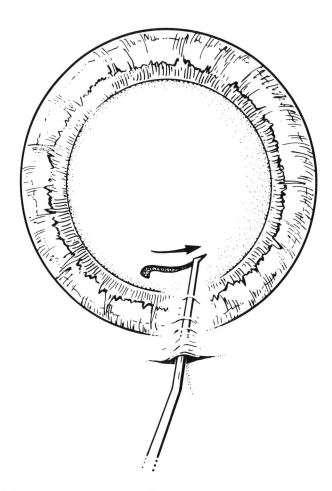

Figure 17–1. Initiating the mini-capsulorhexis. The cystotome pierces the anterior capsule, and a linear continuous tear is created by a gentle sweeping action of the cystotome. The initial capsulotomy is located near the superior iris border in a well-dilated pupil.

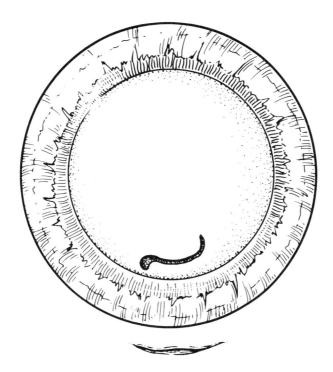

Figure 17–2. The lateral margin of the capsulotomy is extended inferiorly approximately 0.2 to 0.5 mm.

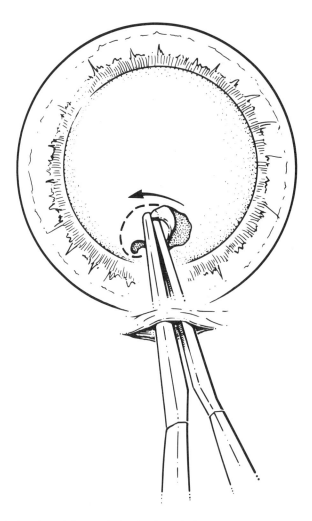

Figure 17–3. Completing the mini-capsulorhexis. Capsule forceps provide best control for completing the mini-capsulorhexis at the superior iris border. The forceps are introduced through a wound enlarged to accommodate the phaco needle. The leading edge of the anterior capsular flap is grasped with the forceps. After completing the inferior margin of the capsulorhexis, this leading edge is directed superiorly and eventually laterally to encompass the original puncture site. Care must be taken to avoid having the lateral margin tear uncontrollably into the superior zonules.

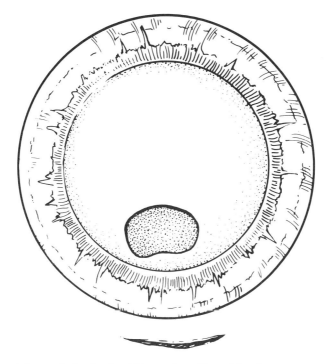

Figure 17–4. The mini-capsulorhexis should be well-positioned to allow easy access for the phaco needle. It should be large enough to accommodate the phaco tip with its irrigating sleeve.

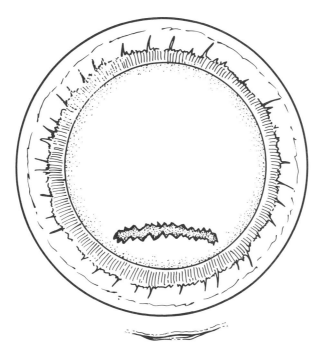

Figure 17–5. A can–opener style capsulotomy creates a series of micro-punctures, any of which poses a significant risk of creating radial tears of the anterior capsule.

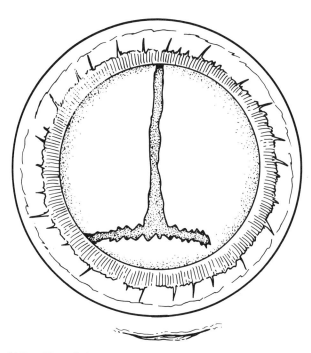

Figure 17–6. About 30% to 50% of phaco cases performed under an anterior capsular flap where the capsulotomy was created by a can–opener technique will result in a split of the anterior capsule. This results in two large anterior capsular flaps. A split in the anterior capsule eliminates many of the protective advantages inherent in the endocapsular technique.

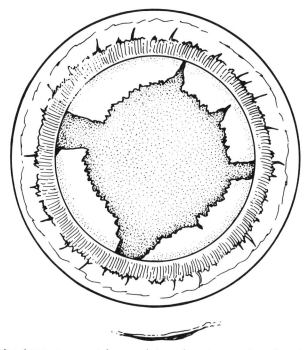

Figure 17–7. Traditional can–opener style capsulotomy in extracaspular cataract extraction following lens expression may result in single or multiple radial extensions of capsular tear into the zonules. Clinically, these tears pose little, if any, risk of extension into the posterior capsule.

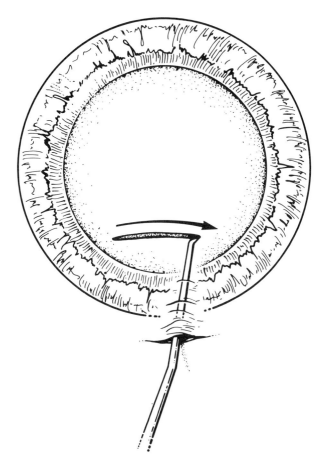

Figure 17–8. The envelope continuous tear capsulotomy. A continuous tear capsulotomy with a smooth edge provides the greatest resistance for preventing the inferior margin of the capsule from rupturing during phacoemulsification. This technique is performed by puncturing the anterior capsule with a cystotome and sweeping in a lateral motion to effectively create a continuous tear capsulotomy near the superior iris border.

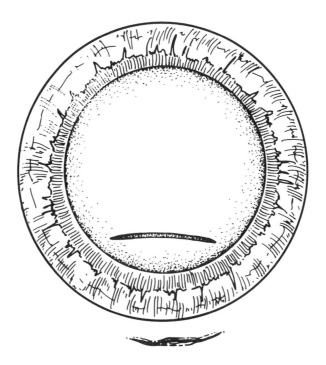

Figure 17–9. The envelope capsulotomy. The envelope capsulotomy results in a continuous tear, allowing a greater percentage of phaco cases to be performed under an intact anterior capsule with minimal risk of rupture to the anterior capsule. It is possible to insert an intraocular lens through this small opening after successful completion of phacoemulsification and cortical removal.

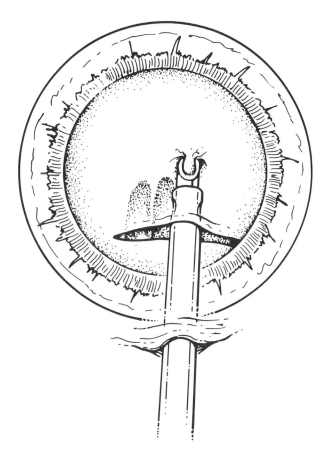

Figure 17–10. The envelope capsulotomy. The drawback of the envelope technique is that stress is applied to the lateral margins of the capsulotomy during phacoemulsification, resulting in superior lateral extension into the zonular area. These generally do not hinder completion of endocapsular phacoemulsification under a relatively intact anterior capsule.

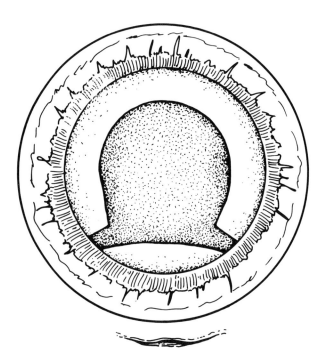

Figure 17–11. The envelope capsulotomy. When the bulk of the anterior capsule is eventually removed, an intraocular lens implanted in the bag will be bound by a capsulotomy of only lateral and inferior margins. Because concentric contraction forces of the capsule exist postoperatively, the intraocular lens may tend to decenter upward following the use of the envelope–style capsulotomy technique.

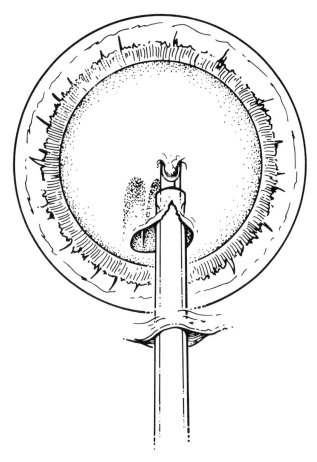

Figure 17–12. Rupture of the inferior margin of the mini-capsulorhexis. Stress from the phaco needle occasionally may result in a small tear of the inferior margin of the mini-capsulorhexis. These tears have the potential of extending radially into the zonules if left unattended.

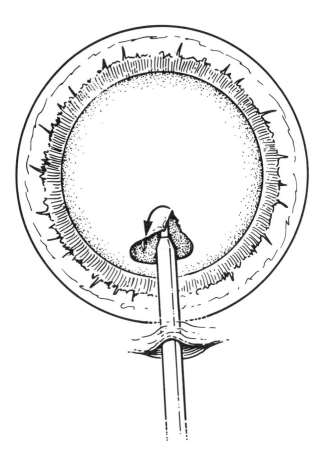

Figure 17–13. Recovery of a small tear in the anterior capsule. Using the I/A tip with the aspiration aperture facing up, the posterior surface of the leading edge of the tear is occluded into the aspirating port, and the leading edge of the tear is directed back to the original capsulotomy.

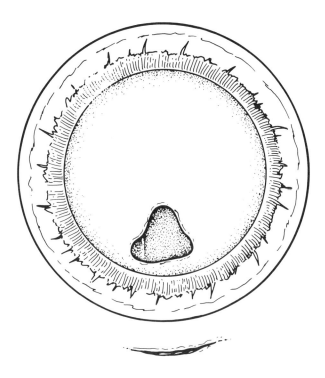

Figure 17–14. Phacoemulsification can be completed through a newly configured capsulotomy that has continuous borders still highly resistant to marginal capsular tears.

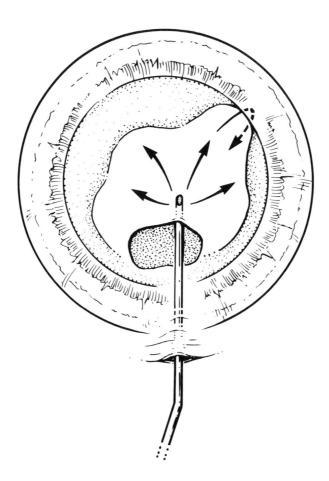

Figure 17–15. Hydrodissection. Hydrodissection is performed prior to emulsification where small volumes (0.2 to 0.5 cc) of balanced salt solution are injected between the cortex and the anterior capsule. This creates a cleavage plane facilitating rotation and dislocation of soft nuclei. This technique is not as useful in a more firm nuclear sclerotic lens.

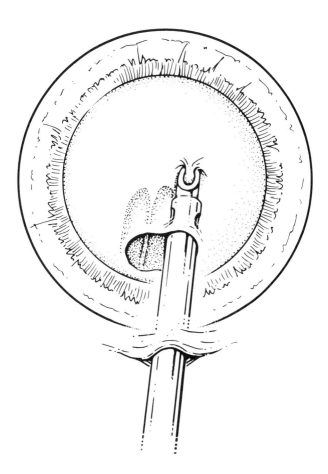

Figure 17–16. Prior to commencing sculpting. Working space for the phaco tip is created with several short bursts of low power linear phaco under the leading edge of the anterior capsule.

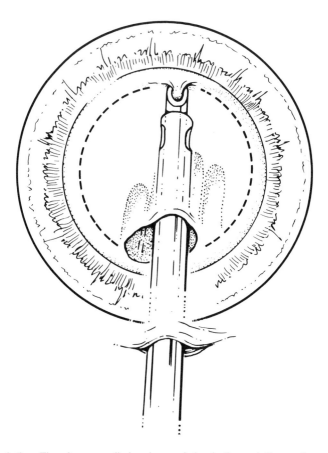

Figure 17–17. Sculpting. The phaco needle is advanced slowly through the nucleus under the anterior capsule, leaving a small rim of superior epinuclear material between the phaco tip and the anterior capsule. The forward progression of the phaco tip in early sculpting should not proceed beyond the zone of the central nucleus and the epinuclear rim.

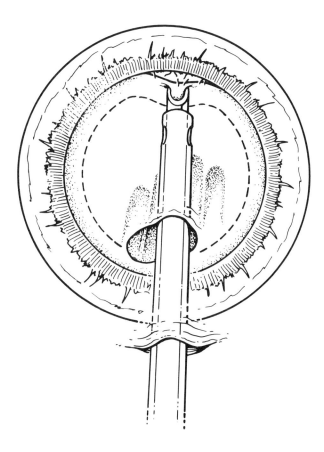

Figure 17–18. The epinuclear rim should be avoided early during the phaco process. This rim is firmly attached to the cortical capsular margins. Premature entry could result in a zonular dehiscence in the affected quadrant.

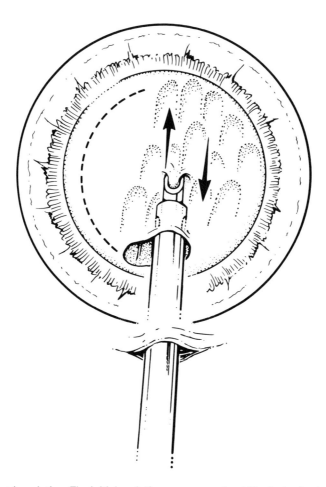

Figure 17–19. Stage I, sculpting. The initial sculpting maneuver should be limited to half of the nucleus, leaving the epinuclear rim intact. The depth of the sculpting should proceed to include at least 2/3 to 3/4 of the central nuclear thickness.

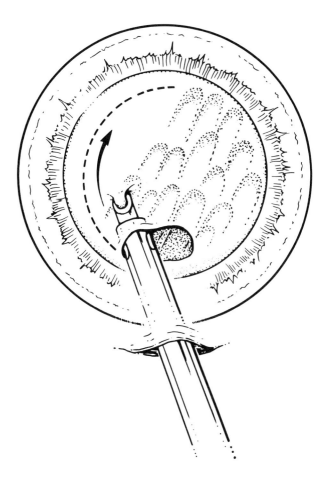

Figure 17–20. Stage II, dislocation and rotation maneuver. The phaco tip is impelled into the base of the superior ledge of the unsculpted rim of the nucleus. Gentle forces are applied with the phaco tip to rotate the nucleus 90 degrees. The unsculpted nucleus is now inferiorly located and ready for further phacoemulsification.

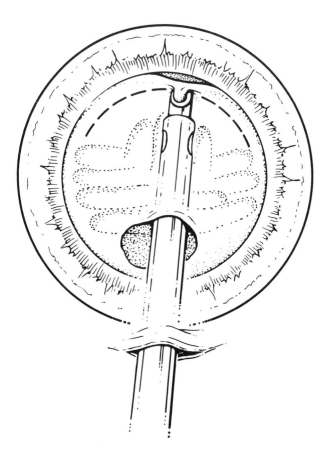

Figure 17–21. Stage III: creation of the nuclear bowl. The remaining central nuclear material now present inferiorly is sculpted, completing the nuclear bowl. The epinuclear rim may be emulsified under low phaco power.

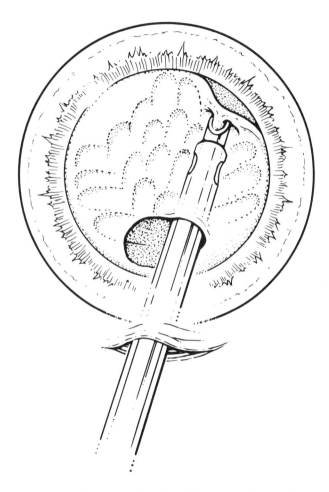

Figure 17–22. Stage III: removal of the epinuclear rim. After the nuclear bowl has been completed, the surgeon can freely emulsify the epinuclear rim. Rotation and dislocation maneuvers should have disassociated any residual cortical capsular attachments from the epinuclear rim. Phacoemulsification under low power poses minimal risks to peripheral capsular engagement.

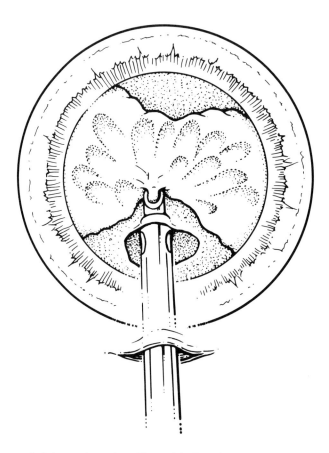

Figure 17–23. Removal of the nuclear plate. The residual nuclear plate is emulsified after it has been dislocated off the posterior capsule and floated anteriorly toward the iris plane.

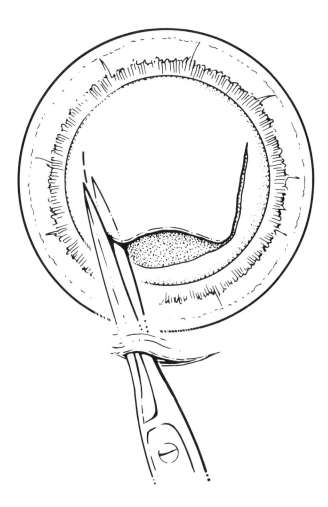

Figure 17–24. Removal of the anterior capsule. The lens capsule is prepared for intraocular lens implantation and the central anterior capsule is removed for optical considerations. A capsule scissor is used to create a small tear in the anterior capsule which is the starting point for a larger second-stage capsulorhexis.

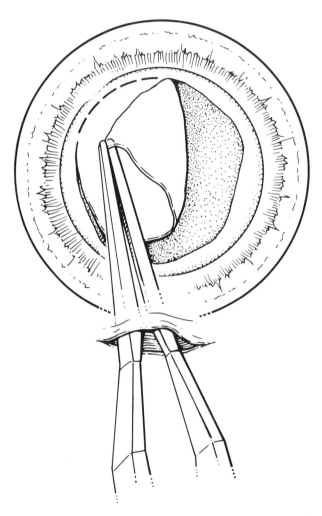

Figure 17–25. Removal of the anterior capsule. The second-stage capsulorhexis is completed with a capsule forceps creating a 360-degree continuous tear capsulotomy.

because the capsular opening will generally be too small to accommodate a standard 6–mm–diameter IOL. Should an IOL be forced through such a small opening, disruption of the superior zonules is likely to occur.

Another important advantage of mini-capsulorhexis is that when the margins of the capsulotomy remain intact without extending to the equator, it can be enlarged to full circular tear capsulotomy prior to IOL insertion. The IOL will be circumscribed by a 360–degree anterior capsular rim, with little risk of clinically significant dislocation.

Hydrodissection Essential to Endocapsular Phaco

The softer and stickier the nucleus, the greater the importance of hydrodissection in endocapsular phacoemulsification.

Primarily, hydrodissection should be used to separate cortical capsular adhesions from the capsule in soft nuclei that exhibit a strong propensity to bind with it. Achieving this separation is extremely important because the lens must be rotated within the lens bag to complete emulsification. Thus, hydrodissection separates cortical capsular adhesions early in the case, greatly facilitating the dislocation of a soft lens during phacoemulsification.

Again, due to the increased likelihood of binding, the softer the lens, the more important hydrodissection becomes; conversely, hydrodissection is less important with harder lenses, as dislocation is considerably easier, and a complete posterior capsule/epinuclear hydrodissection can be done at a later stage using the phaco tip.

To accomplish hydrodissection, the surgeon should inject approximately 0.2 cc's to 0.5 cc's of BSS under the anterior capsule in order to create the cleavage plane described above. This balance salt fluid "wave" should dissect the lens minimally to its equator (Figure 17–15). Creating this cleavage plan eases dislocation of the sticky nucleus from the cortex when rotation of the somewhat "uncooperative" lens with the capsule is attempted. Dislocating the peripheral nucleus also minimizes the possibility of aspirating peripheral capsule into the handpiece.

Should the surgeon not use hydrodissection in patients with soft nuclei, he or she may be faced with phacoemulsifying epinuclear cortical material that is firmly adherent to the capsule. Hence, prior dislocation via hydrodissection significantly reduces the risk of penetrating the posterior capsule with the phaco needle and therefore represents an important step toward risk prevention/management.

Inserting the Phaco Tip

It is very helpful to use a viscoelastic to sustain both anterior depth during capsulotomy and hydrodissection as well as to maintain chamber depth while inserting the phaco tip. After the anterior chamber is filled with viscoelastic, the surgeon may introduce the phaco needle in foot position 0. This will also serve to diminish the rapid extrusion of the viscoelastic from the eye. An added advantage of the endocapsular technique, furthermore, is that since minimal fluid exchange occurs in the anterior chamber during phacoemulsification, the viscoelastic will often be retained throughout a significant portion of the surgery.

After insertion, the surgeon should direct the tip to the capsulotomy opening. Using several short bursts of linear phaco under low power, a working space is created under the leading edge of the anterior capsule. This will facilitate entry of the 3–mm needle and silicone sleeve (Figure 17–16). Once accomplished, the surgeon can begin sculpting under the anterior capsule.

Sculpting

I must stress the vital importance of beginning phacoemulsification in the extreme superior epinuclear cortical zone, slowly shaving the nucleus from top to bottom. While the thickness of the anterior capsule generally makes it resistant to puncture by the phaco needle, this is not always the case. In order to provide added support to the anterior capsule, a thin layer of epinuclear cortical material should be left attached to the undersurface of the anterior capsule during the early sculpting maneuvers. This will help the surgeon prevent an inadvertent rupture of the anterior capsule by the phaco tip. When sculpting is near completion, this material may be removed under low power. However, visualization of the anterior capsule may be more difficult using this approach.

The forward excursion of the phaco needle should extend through the central nucleus, but remain limited to the inner edge of the epinuclear rim (Figure 17–17). In most eyes, this is well before the border of the inferior iris. Care must be taken so that forward excursions of the phaco tip do not violate this circumference, as there is a tendency to phaco through this rim early in the case.

The surgeon should also be aware that it is very important to leave this rim intact in the early stages of sculpting. The rim provides an all-important barrier to protect the equatorial posterior capsule during sculpting, both early and later on during the phacoemulsification process. It should not be removed until the epinuclear cortical attachments have been severed by a rotation/dislocation maneuver (described in Sculpting Section, Stage 2 of this chapter) and after sculpting has been completed.

In order to further visualize the "transition zones" between the central and outer nucleus, we should digress for a moment to draw attention to the added function of hydrodelamination.

Hydrodelamination, the process of injecting a small amount of fluid centrally into the lens to delaminate (separate into layers) the central nuclear zone from the epinuclear zone, serves to establish internal boundaries/guidelines of demarcation to make the surgeon conscious of the peripheral limit of sculpting during phacoemulsification. This assists the surgeon from straying too far peripherally into the epinucleus, which is usually firmly attached to the cortex, which in turn is attached to the capsule. By penetrating the epinuclear zone prior to dislocation of the lens, the capsule may be engaged in the phaco tip, which may disrupt zonules in a specific quadrant (Figure 17–18).

Three-Stage Approach

In Stage 1, phacoemulsification is limited to half of the nucleus. The excursions of the phaco tip are predominantly from the 12 o'clock to 6 o'clock position in order to minimize any lateral movement. A 5- to 10-degree variation is tolerable from the pivot point at which the phaco needle enters the eye, but the basic maneuver should be of a to-and-fro motion and not one of random excursions. After about half of the nucleus has been sculpted to a depth of two phaco diameters, the surgeon is ready for Stage 2 (Figure 17–19).

Stage 2 requires a 90–degree in-the-bag rotation of the nucleus with the phaco tip. The purpose of sculpting only half of the nucleus now becomes apparent. The phaco tip, in foot position 2, is retracted near the mouth of the capsulotomy and engages the base of the nuclear rim superiorly. The lens is rotated in an arc approximately two clock hours at a time, until the peripheral cortical attachments loosen, allowing a 90–degree rotation of the lens (Figure 17–20). The lateral half of the nucleus that had not been sculpted is now present inferiorly (Figure 17–21). Once the lens has been dislocated, the surgeon has far more liberty to phaco the epinuclear rim under low phaco power with less risk of capsular adhesion or rupture, because the rim is free and rotatable within the bag.

Stage 3 involves completing the sculpting of the central nucleus, thus creating the nuclear bowl followed by removal of the epinuclear rim. The phaco needle is then used to sculpt the remaining half of the central nucleus with the same to-and-fro motion. After the nuclear bowl has been created, the surgeon has the liberty to extend the sculpting and emulsification from the central nucleus to emulsify the epinuclear zones (Figure 17–22).

Here, two points should be borne in mind. First, each time the phaco needle is retracted prior to a new sculpt, the surgeon should be sure that the edge of the silicone sleeve does not exit the lens bag. If it does, one should ensure that it clears the edge of the anterior capsulotomy when inserted. If the phaco tip is advanced to the 6 o'clock position and the silicone sleeve has engaged the anterior capsule, a superior zonular dehiscence or tear in the anterior capsule may result.

Second, the surgeon should maintain no greater than a 10- to 15-degree angular pivot with the phaco tip in order to avoid putting stress on the lateral margins of the capsulotomy. This reduces the incidence of linear capsular rents to the equatorial zone from the lateral margins of the anterior capsulotomy.

Removing the Nuclear Plate

The residual epinuclear rim and the nuclear plate should now be removed with the phacoemulsifier. The technique of removal depends upon the firmness of the epinuclear rim. Most of the time, the residual epinuclear rim is gelatinous. It can thus be manipulated off the posterior capsule and emulsified close to the iris plane while under the anterior capsule. However, in cases in which the epinuclear rim is hard, resulting in poor "followability," the inner aspect of the rim should be emulsified segmentally under very low phaco power, using the tip to rotate the nuclear bowl one to two clock hours at a time until the rim has been emulsified 360 degrees. Once the final nuclear plate has been created, it must be lifted off the posterior capsule before emulsification is completed (Figure 17–23).

Proper use of linear power is critically important to the surgeon during sculpting. Linear power facilitates emulsification of the lens near the capsule, both superiorly and peripherally, which helps diminish the risk of capsular rupture. A disciplined surgeon will know the limits of his or her phaco technique when creating the inferior epinuclear rim and will use low power to obtain it.

The major hazard to be addressed is that of not properly controlling the phaco power. Phacoing the nuclear rim under inappropriately high power may result in a sudden surge of the rim into the phaco needle while the rim is still attached to the peripheral capsule, resulting in a peripheral capsular penetration of the phaco tip. This situation, however, is completely avoidable. Knowing how to use minimal power (in the range of 10% to 30%), while carefully controlling the foot position back into position 2, will give the disciplined surgeon superior control while creating the inferior rim, without the associated risk of capsular penetration.

Cortex Removal

Cortical removal is easily performed under a relatively intact anterior capsule, with irrigation contained, as previously described, within the lens bag. Again, the presence of the anterior capsule provides excellent protection to the corneal endothelium during I/A.

Removal of Anterior Capsule

Prior to removing the central anterior capsule, it is important to remove the peripheral lens element from the anterior capsule; if desired, the lens epithelial cells may also be removed.[5,6]

After completion of capsule polishing, the anterior capsule is ready to be removed. Capsule scissors should be used to make a small cut into the capsule (Figure 17–24). Following this, capsule forceps may be employed to complete the "second stage" of the capsulorhexis, thereby creating a larger capsulotomy to accommodate the intraocular lens in the capsular bag (Figure 17–25).

Conclusion

Endocapsular Phacoemulsification is a technique that I believe provides the surgeon with an entry point into the future of ophthalmology and cataract surgery, given the research now taking place in the fields of injectable intraocular lenses, endocapsular balloons, and many other groundbreaking technologies.[7,8,9]

References

1. Patel J, Apple D, Hansem S, Solomon K, Tetz M, Gwin T, O'Morochoe D, Daun M: Protective Effect of the Anterior Lens Capsule During Extracapsular Cataract Extraction. Part II. Preliminary Results of Clinical Study. Ophthalmology 96: 598–602, 1989.

2. Solomon K, Todd G, O'Morochoe D, Tetz M, Hansen S, Sugita A, Imkamp E, Apple D: Protective Effect of the Anterior Lens Capsule During Extracapsular Cataract Extraction. Part I. Experimental Animal Study. Ophthalmology 96: 591–597, 1989.

3. Hara T, Hara T: Endocapsular Phacoemulsification and Aspiration (ECPEA)—Recent Surgical Technique and Clinical Results. Ophthalmic Surg 20: 469–475, 1989.

4. Assia E, Apple D: Presented ACSRS Los Angeles, 1990, personal communication.

5. Nishi O: Lens Epithelial Cell Removal by Ultrasound: Access to 12 o'clock. J Cataract Refract Surg 15: 704–706, 1989.

6. Nishi O: Intercapsular Cataract Surgery with Lens Epithelial Cell Removal. Part I: Without Capsulorhexis. J Cataract Refract Surg 15: 297–300, 1989.

7. Nishi O: Refilling the Lens of the Rabbit Eye After Intercapsular Cataract Surgery Using Endocapsular Balloon and an Anterior Capsule Suturing Technique. J Cataract Refract Surg 15: 584–588, 1989.

8. Nishi O, Hara T, Hara T, Hayashi F, Sakka Y, Iwata S: Further Development of Experimental Techniques for Refilling the Lens of Animal Eyes with a Balloon. J Cataract Refract Surg 15: 584–588, 1989.

9. Hara T, Hara T: Recent Advances in Intracapsular Phacoemulsification and Complete in-the-bag Intraocular Lens Implantation. Am Intra-Ocular Implant Soc J 11: 488–490, 1985.

Section IX

CONCLUDING THE OPERATION

18

Small Capsulotomy Cortical Aspiration Techniques

Paul S. Koch, MD & James A. Davison, MD

For the most part, cortical aspiration following an endocapsular procedure is easy. A lot of the cortex was removed during the emulsification of the nucleus. What's left has been hydrated by all of the fluid irrigated into the capsular bag and is easily removed.

During the cortical aspiration in a capsulorhexis case, the anterior capsule remains taut and intact, stretched out in its original plane, so it's easy to reach under it to get the cortex. There are no loose capsular flaps that can be aspirated into the aspiration port, interfering with the cortical removal.

There is a tendency to relax during this relatively easy step after phacoemulsification. It is still easy to tear a posterior capsule.

The problem comes when trying to aspirate the sub-incisional cortex. That's a pretty hard step in can-opener capsulotomy phacoemulsification techniques because it's hard to get around that sub-incisional corner. But at least the anterior capsule will give way and give you a chance. In capsulorhexis phacoemulsification, the anterior capsule tries to stay taut and intact, preventing you from getting the aspiration tip around it to where the cortex is sequestered.

There are three ways to get the sub-incisional cortex.

1. Direct Aspiration

Many times the first step in I/A is to remove some small nuclear fragments (Figure 18–1). After this is accomplished, the superior cortex should be removed. The handpiece must be relatively more vertical to effectively apply the 0.3 mm aperture to the cortical fibers, which are adherent to the anterior capsular remnant at 11 o'clock (Figures 18–2, 18–3). Some of the protruding fibers will be caught either at 11 o'clock or on either side of it. Anterior and equatorial

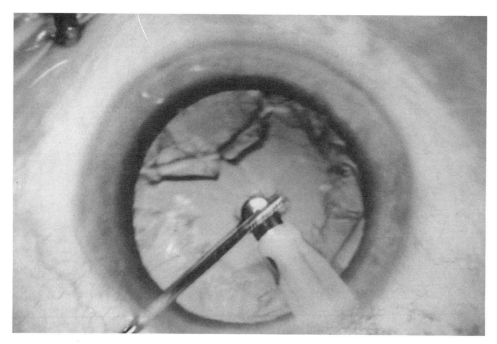

Figure 18–1. Aspiration often starts by removing one or two small nuclear fragments and crushing them into the aspiration tip with the cyclodialysis spatula.

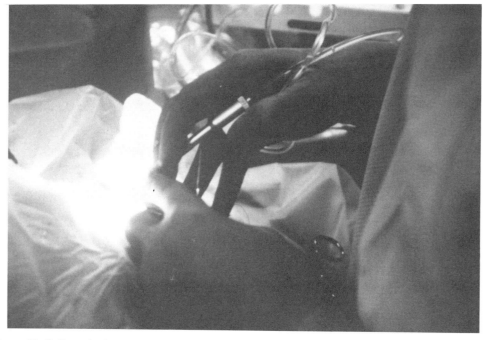

Figure 18–2. Note the hand position required for superior removal after capsulorhexis. Observe the steep attitude required of the I/A handpiece.

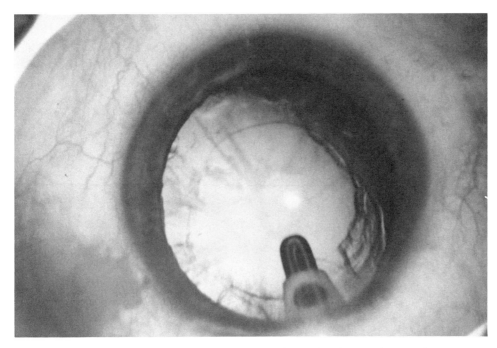

Figure 18-3. The 11 o'clock cortex is grasped at its edge just under the anterior capsular remnant.

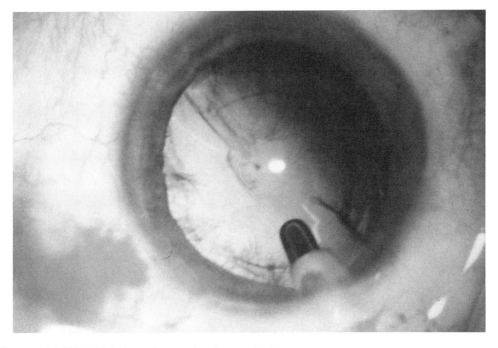

Figure 18-4. This initial piece of cortex has been pulled away.

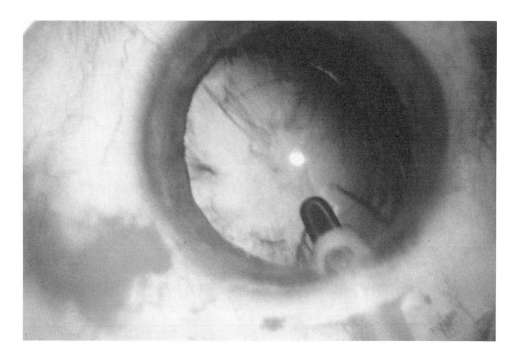

Figure 18–5. Adjacent cortex is grasped again at the anterior capsular edge.

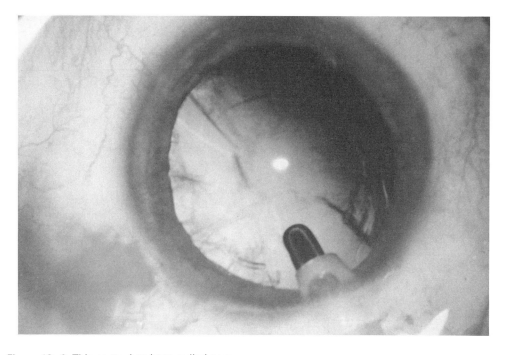

Figure 18–6. This cortex has been pulled away.

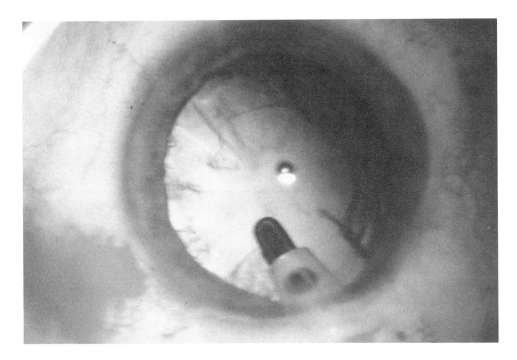

Figure 18-7. Another small adjacent strip is grasped and withdrawn.

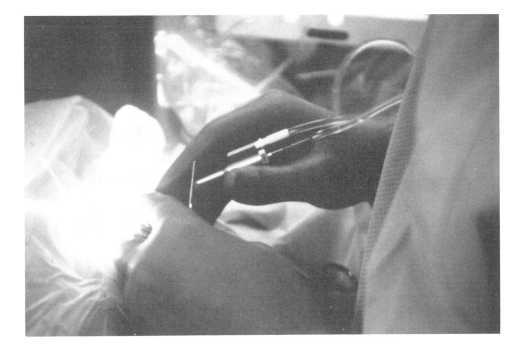

Figure 18-8. After the initial superior cortex has been removed, the hand position automatically flattens to accomplish removal of the inferior and right and left sides.

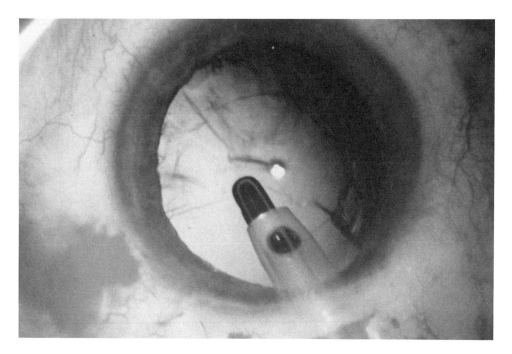

Figure 18–9. Cortex removal is accomplished on the left side.

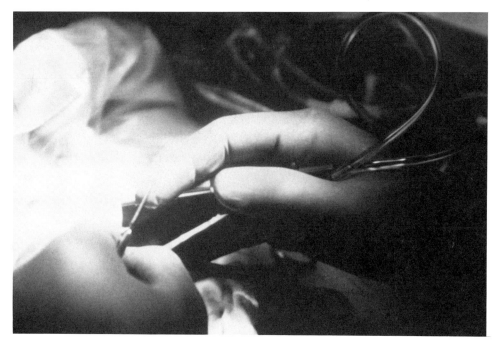

Figure 18–10. Hand position changes for left sided cortex removal. Some fingers still rest on the forehead to achieve good stabilization.

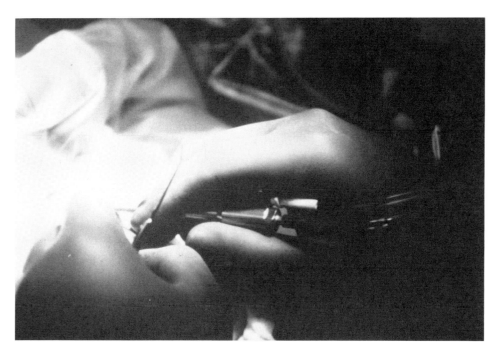

Figure 18–11. New hand position is present. Notice that both hands still work together holding the I/A handpiece.

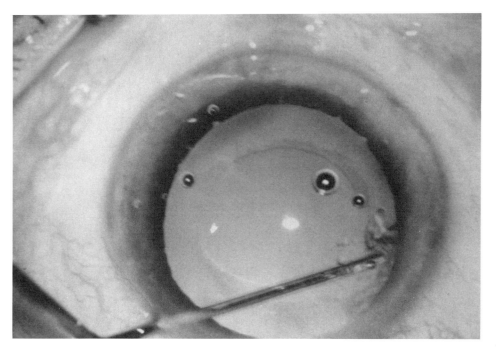

Figure 18–12. After some viscoelastic is placed, the cortex is roughed up with a cyclodialysis spatula in an attempt to get an edge to grasp.

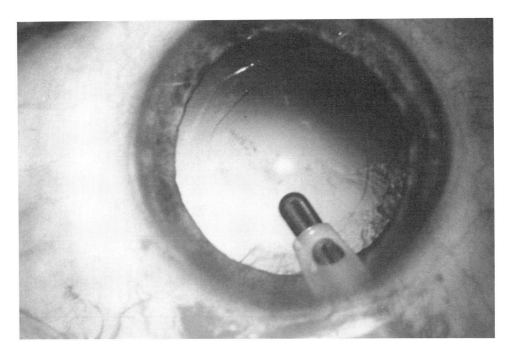

Figure 18–13. The problem of superior cortex removal can be more troublesome in filmy posterior capsular cataracts than in other types. Sometimes the cortex just has to be left.

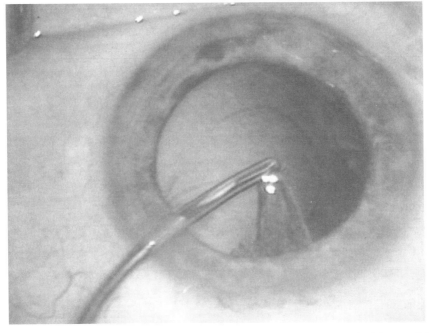

Figure 18–14. Sideport cannulation. A Simcoe cannula is placed in the sideport incision, and the sub-incisional cortex is aspirated manually. This can be facilitated by rotating the eye upward.

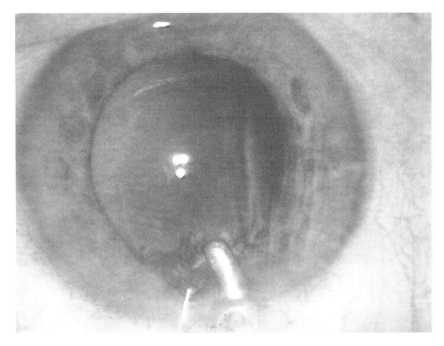

Figure 18–15. Capsular retraction. Open the incision to the full length needed for the implant. Use a hook to retract the iris and capsule, facilitating exposure of the sub–incisional cortex.

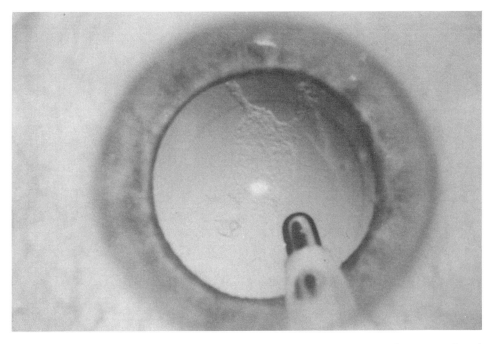

Figure 18–16. The 0.2-mm I/A tip carefully vacuums the posterior capsule. Intermittent taps of suction are used as the tip wags its way back and forth. The infusion bottle may be lowered to 30 cm—40 cm above the eye if the capsule is pushed too far posterior by infusion pressure.

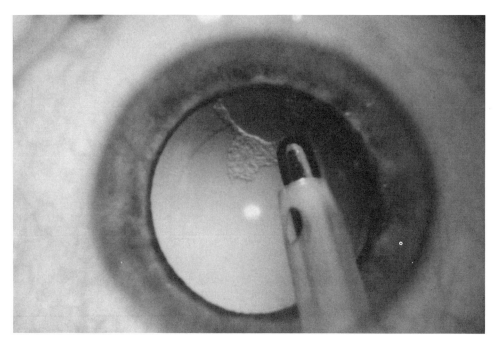

Figure 18–17. Most of the capsule has been vacuumed.

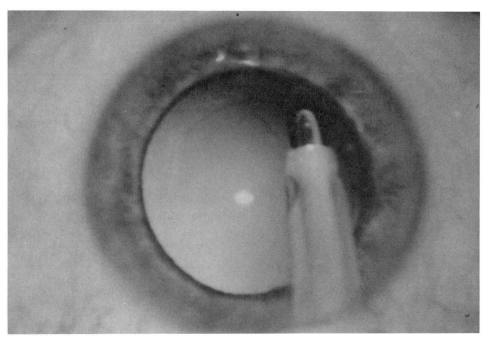

Figure 18–18. A small amount of peripheral debris remains. This will be ignored because of the risk to the anterior capsular remnant.

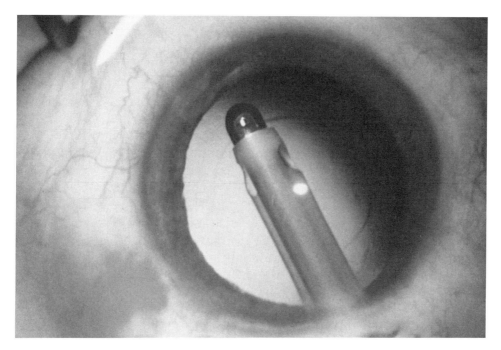

Figure 18–19. Anterior capsule vacuuming may decrease epithelial cell complications.

cortex will strip away without difficulty when starting above and moving peripherally (Figures 18–4 through 18–11). If the cortex is removed progressing from inferior to superior, support and presentation is lost at the 11 o'clock position. Even though an angled I/A tip may be used, some cortex may have to be left behind (Figures 18–12, 18–13).

It may be possible during cortical aspiration to aspirate a portion of the cortex somewhere near the sub-incisional area and use this piece to help tease the sub-incisional cortex out of the fornix and into the capsular bag. The aspiratable portion sometimes is a central piece of the sub-incisional cortex, and other times it may be a piece to one side or another. Provided the portion of cortex is not hit with high aspiration and gobbled up, it can be used as a wand to pull sub-incisional cortex into a more accessible zone.

2. Sideport Cannulation

If direct aspiration is not possible, sideport cannulation may be easily performed. The anterior chamber is filled with a viscoelastic to prevent chamber collapse during aspiration. Then, a curved Simcoe cortical aspiration cannula placed on a 3–cc syringe filled with BSS is passed through the sideport incision, 90 degrees away from the sub-incisional area. The aspirating tip is placed under the capsule and the sub-incisional cortex is aspirated gently and teased into the center of the capsular bag (Figure 18–14). There, it may be aspirated with either the Simcoe cannula or the I/A handpiece. The preferred method depends on how much cortex there is. A little can be removed with the manual technique, but a lot of cortex is more easily removed with the I/A handpiece.

3. Capsular Retraction

If there is a lot of sub-incisional cortex, the sideport cannulation method may take a lot of work and repetition. In this case, it may be helpful to use capsular retraction.

One way to do this is to place a blunt or forked instrument through the sideport incision and push the iris and capsule toward the incision, opening up a space for the I/A handpiece to get around the corner and aspirate the sub-incisional cortex. This is most appropriate if the incision will be no more than 4 mm after the lens is inserted.

The second way to do capsular retraction is to open the incision to the length needed for the lens implant. Place the I/A handpiece through one half of this incision and, place an iris hook of one sort or another through the other half of the same incision. Retract the iris and capsule with the hook and aspirate the cortex with the I/A handpiece (Figure 18–15).

Vacuuming the Posterior Capsule

The irrigation bottle is lowered to 40 cm above the eye, and the machine placed in a capsular vacuum mode. A 0.2 mm tip is used to vacuum the posterior capsule as well as the conveniently available anterior capsular remnant (Figures 18–16 through 18–19)

19

Technique for Intercapsular Phacoemulsification With Two-Staged Capsulorhexis

Jack A. Singer, MD

New surgical techniques, including capsulorhexis, hydrodissection, and hydrodelineation, have made it possible to safely and reproducibly perform phacoemulsification between the anterior and posterior capsules. Therefore, having experience with these techniques will greatly reduce the learning curve of intercapsular phacoemulsification.

Pre-Operative Consideration for Intercapsular Phacoemulsification

1. Anesthesia—A retrobulbar block may be preferable to a peribulbar block for intercapsular surgery, because a lower volume of anesthetic is required in the orbit.
2. Orbital and Ocular Compression—A Honan cuff, mercury bag, or similar device should be applied following the retrobulbar block for a minimum of twenty minutes. This greatly reduces the chance of positive vitreous pressure interfering with maneuvers within the capsular bag.
3. Lid Retraction—I routinely use Steri-strips for lid retraction, which eliminates the need for a lid speculum. Lid speculums frequently produce positive vitreous pressure by exerting pressure on the globe.
4. Maximizing Intraocular Visualization—The use of a fluid wick in the inferior conjunctival fornix helps to prevent pooling of fluid on the cornea, which can impair visualization during emulsification. This is par-

ticularly helpful in deep-set eyes. A thin coating of viscoelastic on the epithelium, moistened with a few drops of BSS, usually is sufficient to maintain corneal epithelial clarity during the entire procedure. This eliminates the need to constantly re-wet the cornea.

Sideport Incision

A 1 mm sideport incision is made at the one o'clock surgical limbus with a 15–degree blade. Although the technique described here for intercapsular phacoemulsification does not routinely employ a nucleus manipulator, it is wise to create a sideport incision at the start for cases in which conversion to a two-handed technique is necessary.

First Stage of Capsulorhexis

Since capsulorhexis has made it possible to safely perform phaco entirely within the capsular bag, the most important element in mastering intercapsular phacoemulsification is the proper use of capsulorhexis. The opening should be small enough to leave an adequate amount of anterior capsule (to provide protection to the corneal endothelium) while also being large enough to allow manipulation of the phaco tip without tearing the anterior capsule. This small opening leaves the capsule much stronger and reduces the number of capsular tears. An initial linear capsular opening is easily made with a 15–degree blade (Figure 19–1). The anterior chamber is filled with a viscoelastic. Using forceps, a horizontal, oval, continuous tear capsulotomy is created in the superior 25% of the dilated pupillary area (Figures 19–2 and 19–3). The dimensions of the opening are approximately 1.5 mm x 4 mm. As with any capsulorhexis technique, it is very important to blend the tear at its conclusion from the outside inward in order to prevent any discontinuities in the capsular opening.

Another technique, termed "button hole anterior capsulotomy" for intercapsular surgery, described by Okihiro Nishi, M.D., involves a linear slit with two circles at each end. The circles can be created with a capsulorhexis technique or a round micro punch, which Dr. Nishi has designed.

Anterior Hydrodissection and Hydrodelineation

1. Anterior hydrodissection—The first step is to form a fluid cleavage plane, separating the anterior capsule from its cortical attachments with an injection of BSS through a blunt 30–gauge cannula. Further injection then separates the outer nucleus from the cortex, leaving some posterior cortical attachments to hold the nucleus stationary during the initial phase of the emulsification (Figures 19–4, 19–5).
2. Hydrodelineation—The cannula is advanced into the outer nucleus until minimal resistance is felt; then, using an additional injection of BSS, the nucleus is separated into two concentric zones: an outer softer zone and

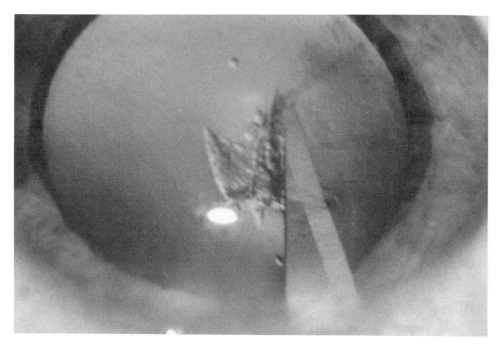

Figure 19–1. A linear opening in the anterior capsule is made with a 15–degree blade.

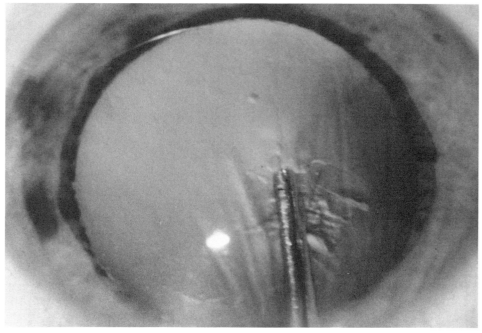

Figure 19–2. The inferior edge of the linear capsule opening is picked up with the forceps to begin the first stage of capsulorhexis.

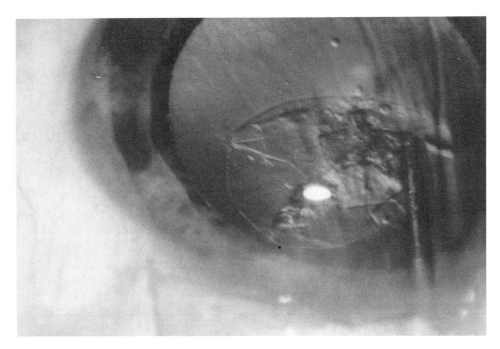

Figure 19–3. Continuous oval capsulorhexis using Utrata capsule forceps to create a horizontal oval continuous tear capsulotomy in the superior 25% to 30% of the pupillary area.

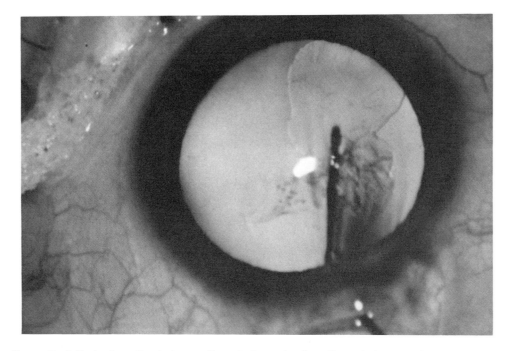

Figure 19–4. Hydrodissection to loosen the anterior cortex from the anterior capsule.

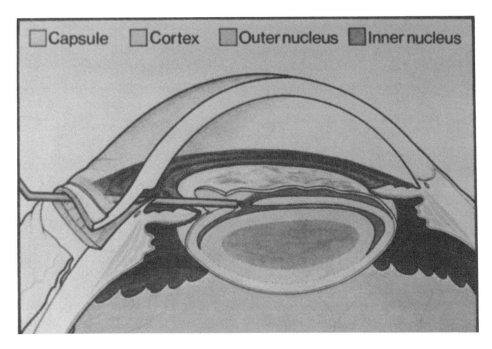

Figure 19–5. Hydrodissection between anterior capsule and cortex and between cortex and anterior outer nucleus (done through a continuous oval capsulorhexis).

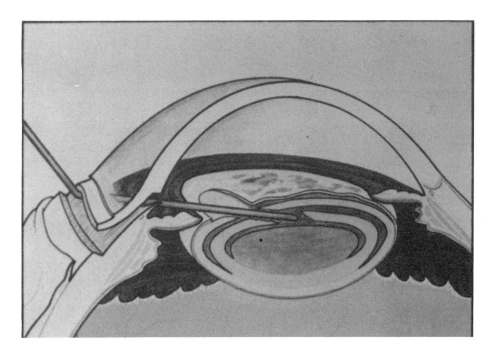

Figure 19–6. Hydrodelineation between inner and outer nuclear zones to create the demarcation zone (DMZ).

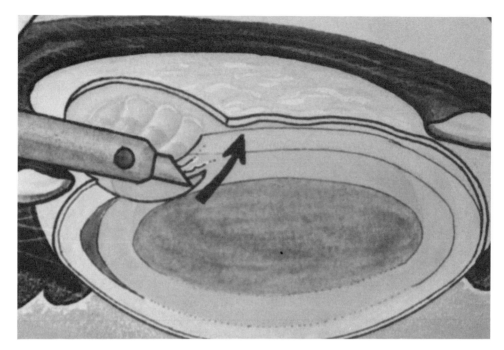

Figure 19-7. Initial sculpting of outer nucleus in the area of the capsulotomy.

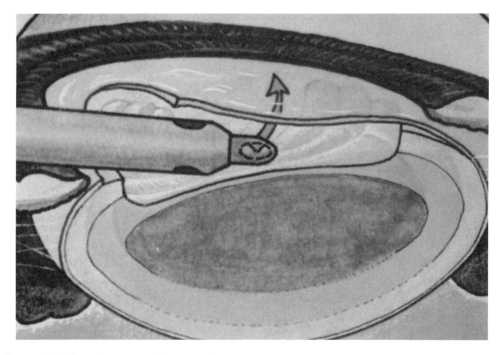

Figure 19-8. Anterior outer nucleus sculpting beneath the anterior capsule.

Figure 19–9. Inner nucleus sculpting.

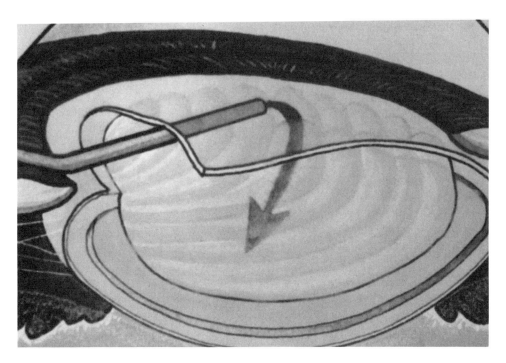

Figure 19–10. Posterior hydrodissection between the outer nuclear zone and the posterior cortex.

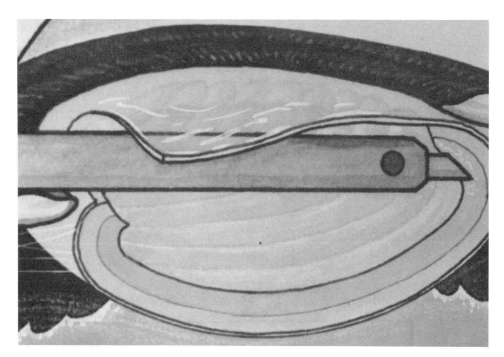

Figure 19–11. Outer nuclear rim removal. Sandwiching of rim to occlude phaco tip.

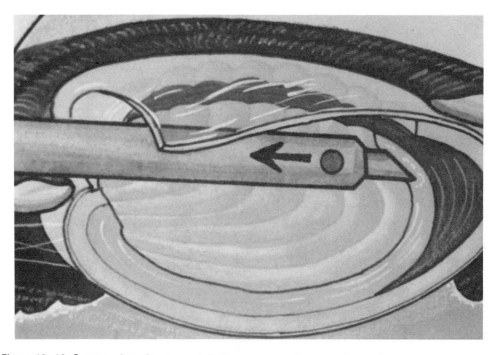

Figure 19–12. Outer nuclear rim removal. Pulling rim toward center of capsular bag.

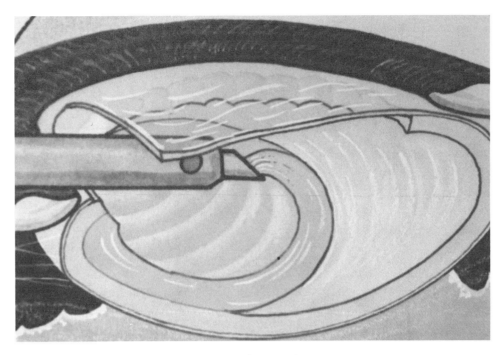

Figure 19–13. Outer nuclear rim removal. Emulsification of rim.

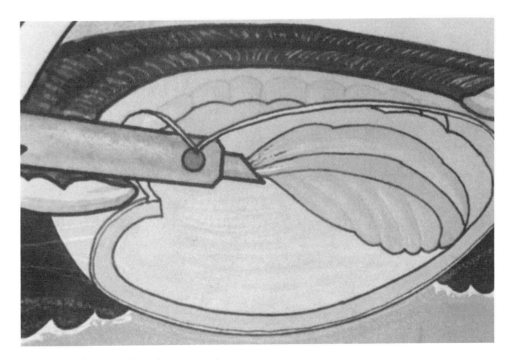

Figure 19–14. Outer nuclear plate removal.

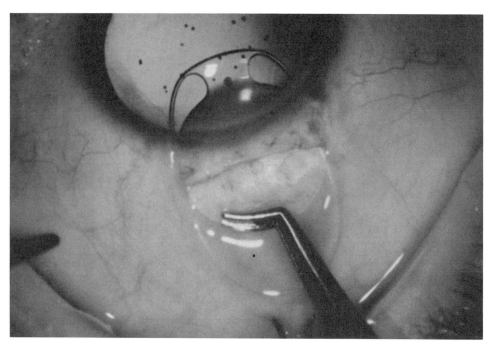

Figure 19–15. Insertion of a Domilens Model Chip-2, one–piece PMMA IOL with a 7–mm circular optic connected to a 9.75–mm circular haptic.

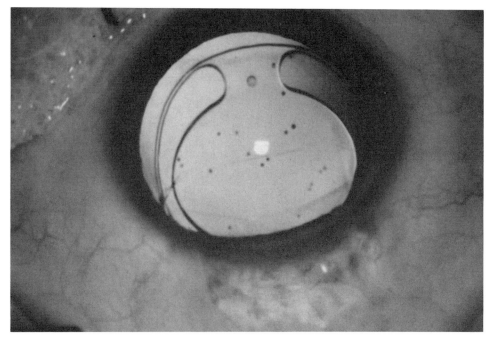

Figure 19–16. The oval continuous tear capsulotomy stretches to allow insertion of the implant into the capsular bag.

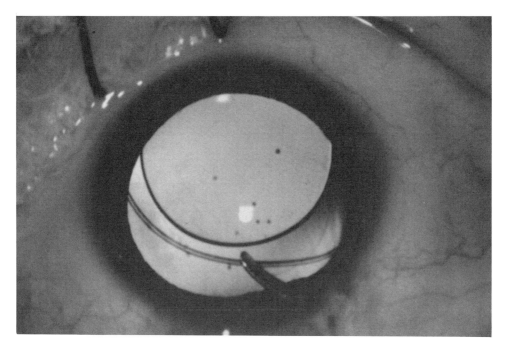

Figure 19–17. The remainder of the circular loop is flexed into the capsular bag.

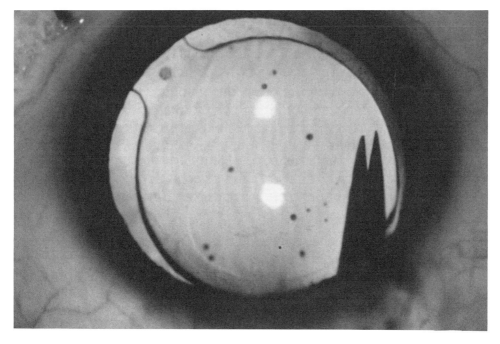

Figure 19–18. A snip is made at the edge of the oval continuous tear capsulotomy with Vannass scissors.

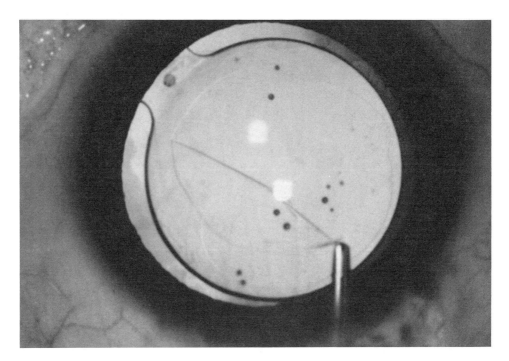

Figure 19–19. A continuous circular capsulorhexis is performed with forceps to enlarge the capsulotomy over the lens optic, maintaining complete enclosure of the lens optic

an inner harder zone. Usually at this time, two concentric circles can be seen through the operating microscope, outlining the separation between the outer nucleus and the cortex and the split between the two nuclear zones, which is known as the demarcation zone (DMZ). Using hydrodissection and hydrodelineation in this manner results in the cataract being divided into two concentric zones and isolated from the surrounding cortex and capsule. This approach gives the surgeon a visual guide during emulsification, which greatly reduces the chance of inadvertent capsular contact with the phaco tip (Figure 19–6).

Equipment and Parameter Setting

I prefer a large bore 45–degree phaco tip, with a diameter of 1.2 mm that decreases phaco time and allows deep sculpting and cutting without unwanted occlusion of the phaco tip. Linear control of ultrasound power with a full range is used during the entire procedure. A flow rate of 12 cc to 16 cc per minute and suction pressure of 80 mm to 105 mm of mercury are employed. The irrigating fluid bottle height is placed at the lowest level to maintain adequate anterior chamber depth and adequate expansion of the capsular bag. The use of a machine with fluid venting and automatic surge prevention helps to improve safety when working within small confines of the capsular bag. Also, audio feedback of suction levels helps improve control of intraocular maneuvers, especially for the beginning surgeon.

Nucleus Emulsification

Although each case is unique and many cases will not require all the steps listed here, particularly soft and very hard nuclei, I have found the majority of cases to involve the following steps:

1. Initial sculpting—The outer nucleus is sculpted within the area of the oval capsulorhexis (Figure 19–7).
2. Anterior outer nuclear sculpt—The anterior portion of the outer nucleus is sculpted to create a working space beneath the anterior capsule. This is best achieved with the bevel of the phaco tip facing sideways to prevent occlusion of the phaco tip with the anterior capsule (Figure 19–8).
3. Inner nucleus removal—Working within the demarcation zone, the dense inner nuclear zone can be safely sculpted while the outer nuclear zone acts as a protective layer to protect and cushion the posterior capsule (Figure 19–9).
4. Posterior hydrodissection—If the remaining outer nucleus does not rotate freely, additional hydrodissection is performed until the outer nucleus is completely separated from the posterior cortex. The tip of the cannula is placed between the anterior capsule and the equator of the outer nucleus. In some cases, the outer nuclear rim can be fractured inward with the cannula while simultaneously injecting BSS. This additional step of posterior hydrodissection greatly facilitates the remaining steps of the emulsification (Figure 19–10).
5. Outer nuclear rim removal—A portion of the outer nuclear rim is removed between five o'clock and seven o'clock in the following manner. With the bevel facing upward, the phaco tip is advanced into the outer nuclear rim, which can then be sandwiched between the phaco tip and the anterior capsule to help occlude the phaco tip. The anterior capsule acts as a second instrument. This creates suction to hold that portion of the outer nuclear rim, which can then be drawn toward the center of the capsular bag. The nucleus is rotated with the phaco tip without irrigation in order to position a fresh portion of the outer nuclear rim between the five o'clock and seven o'clock position. The above manner is then repeated until the majority of the outer nuclear rim is removed (Figures 19–11 through 19–13).
6. Outer nuclear plate removal—The remaining nuclear plate, which is usually small, can easily be emulsified and aspirated using pulsed ultrasound power (Figure 19–14).

Cortex Aspiration

Cortex aspiration can be performed in the usual manner with an 0.3 mm I/A tip. One should, however, keep the aspiration port away from the anterior capsule during the cortex removal to prevent inadvertent occlusion of the tip with the anterior capsule.

Capsule Vacuuming

The broad remnant of anterior capsule is easily vacuumed with an 0.3 mm I/A tip using high suction settings in order to remove as much of the lens epithelial cells as possible. The superior portion of the anterior capsule can be vacuumed with a U-shaped aspiration cannula. The posterior capsule is then vacuumed in the standard manner, using very low vacuum pressures.

IOL Insertion

I prefer to implant the IOL prior to enlarging the capsulorhexis. The small oval continuous tear capsulorhexis is remarkably elastic and stretches as the implant is inserted into the capsular bag. It is almost like inserting the implant into a pocket; when inserted at this stage of the procedure, capsular fixation is virtually assured and easily visualized (Figures 19–15 through 19–17).

Second Stage of Capsulorhexis

Once the implant is safely within the capsular bag, a snip is made at one edge of the horizontal oval continuous tear capsulotomy with Vannass scissors (Figure 19–18). Using forceps, a continuous circular capsulorhexis is performed to enlarge the capsular opening over the visual axis to a diameter that maintains complete enclosure of the lens optic as well as long-term stability and centration of the implant. The presence of the implant prior to the second stage of the capsulorhexis acts as a visual guide to the size of the capsular opening needed (Figure 19–19).

Summary

The protective effect of the anterior capsule on the corneal endothelium has been demonstrated in controlled clinical studies by Jay Patel, MD and David Apple, MD. (**Ophthalmology**. 1989; 96:598–602). ICPE is my technique of choice and is well suited to cases where the endothelium is at risk. It improves control over the intraocular environment, reduces trauma to delicate intraocular structures, and allows accurate and stable capsular IOL fixation. Capsulorhexis and hydrodissection, both of which can be easily learned, facilitate ICPE, which need not be more complex than other methods. Surgeons learning ICPE can simply convert to a more conventional technique should problems arise.

20

Small Capsulotomy Intraocular Lenses

Alan S. Crandall, MD

The technique for performing capsulorhexis has been described in Chapters 3 through 5 of this text. When performed correctly, the capsule remains extremely strong and will not tear even if a large (7−mm optic) intraocular lens is inserted through it.

Previous cadaver studies have shown a poor correlation between the surgeon's intention and the actual placement of the IOL haptics. But, with the technique of capsulorhexis, the surgeon can be certain of haptic position (Figures 20−1, 20−2). This should lead to better IOL centration and to better isolation of the IOL from uveal tissue. While these two factors are obvious, long-term studies will be needed to verify that they are clinically significant. Immediate positive results have been seen in centration of the lenses and decreased glare. The ability to be certain of centration may allow for the placement of IOLs with slightly smaller optics.

Anatomic studies have shown that the capsule is 9.6 +/− 0.4 mm in diameter. Since placement within the capsular bag can now be assured, lens design can be altered. Lenses that are 14 mm long will no longer be necessary and may actually put undue stress on the capsule. The lens may not need a large optic, since centration can be assured.

The technique for implanting the lens through a capsulorhexis is different than a regular implantation technique. For three-piece lenses, Kelman-McPherson forceps can still be used to place the superior haptic in the capsular bag.

The all-PMMA lenses can be easily rotated into the capsular bag since the capsular opening will expand and then close over the lens. The inferior haptic is placed under the capsule and slightly rotated counter-clockwise. The optic is placed into the capsular bag by pushing slightly inferiorly and toward the optic nerve. If the inferior haptic is not under the capsule, it will "pop" up. A Sinsky-type hook is used to rotate the superior haptic into the capsular bag.

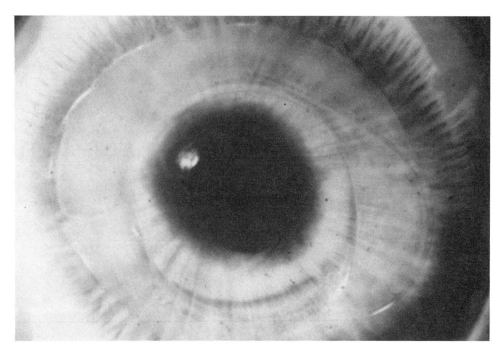

Figure 20–1. A rabbit eye eight weeks after implantation of an IOL by the capsulorhexis technique.

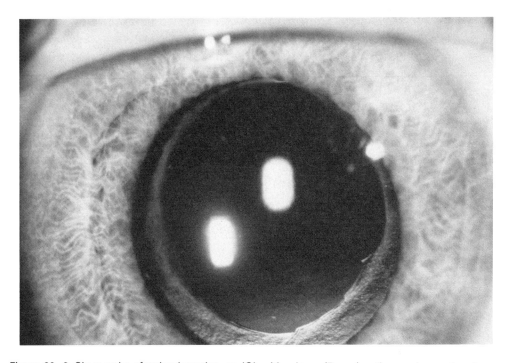

Figure 20–2. Six months after implantation, an IOL with a large (7 mm) optic remains centered.

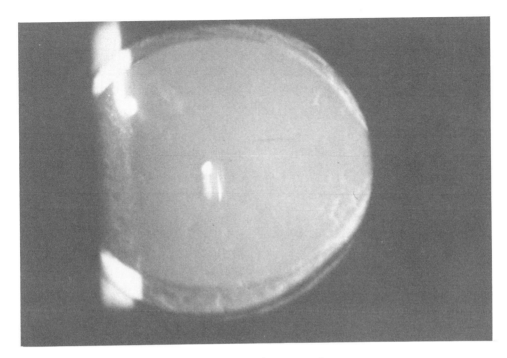

Figure 20–3. A large (6.5 mm) all-PMMA lens remains centered.

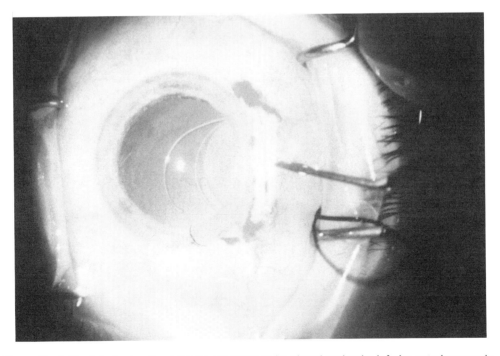

Figure 20–4. The inferior haptic of the lens is about to be placed under the inferior anterior capsule.

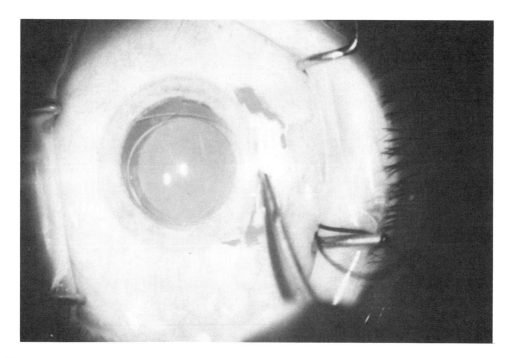

Figure 20–5. Begin rotation by bringing the 3 o'clock haptic into the capsular bag.

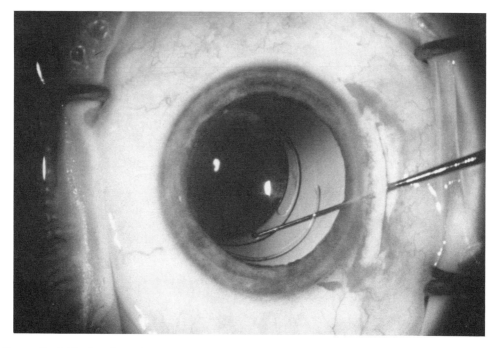

Figure 20–6. The lens is pushed toward 6 o'clock, and the capsule slides over the lens at 9 o'clock.

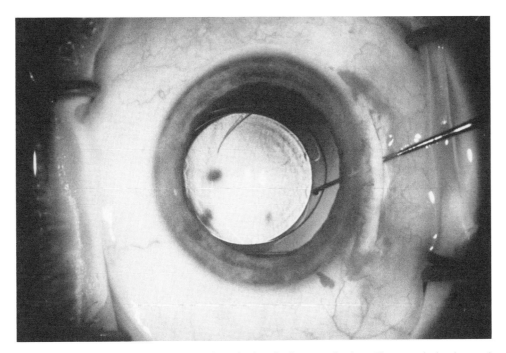

Figure 20–7. The Sinsky hook is used to place the lens in the capsular bag. The capsulorhexis opening is stretched open to accommodate it.

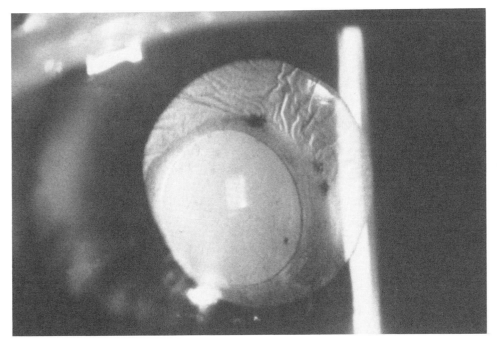

Figure 20–8. A Staar lens shows an intact superior capsule after the Graether collar-button technique is used to stretch the capsular bag opening for placement

The optic is dropped into the posterior chamber and a clockwise rotation begun. The optic will not be moved out of the center of the posterior chamber (Figure 20–3).

The "slab" style IOL (such as those made by Staar and Iogel) can be placed easily, because the capsule opening is tough enough to withstand pressure from the superior haptic and the surgeon can easily fold the haptic into the capsular bag. The opening is also strong enough to withstand stretching with a hook or collar button so that the superior portion of the haptic can be placed into the capsule (Figure 20–4).

Newer style lenses, such as disc lenses, smaller IOL's with C-loops, and, of course, foldable lenses, are all being evaluated at this time, and initial results have been very encouraging.

21

Multifocal Intraocular Lenses

Robert P. Lehmann, MD

At the "Innovations in Multifocal Lens Design" symposium held in Copenhagen during the Sixth Congress of the European Intraocular Lens Implant Council, Richard Lindstrom stated that the multifocal lens "is probably the first truly major advance in intraocular lens implantation since Ridley." Lindstrom went on to add that when Ridley introduced his concept of pseudophakia in November 1949, he stated that "the first complication of cataract extraction is aphakia." Lindstrom then added that "perhaps it can be equally stated that the first complication of lens implantation is presbyopia." He predicted that within the next decade, 90% to 95% of patients undergoing cataract removal would receive a multifocal implant of some kind. Having had personal clinical experience with two different implant designs, I agree with this forecast.

Until recently, all available intraocular lenses have been monofocal, or single vision. The surgeon's goal was to provide good distance acuity, often with slight overcorrection of one or both eyes to afford some degree of improved near acuity. Some surgeons advocate a type of monovision for certain patients. This requires overcorrecting the nondominant eye by one to three diopters to allow for enhanced near vision in that eye. Others advocate surgically inducing myopic astigmatism as a viable alternative to multifocal intraocular lenses.

I could not disagree with this strongly enough, as it is our accepted goal to reduce postoperative astigmatism. Is it not an even more demanding task to attempt the incorporation of a fixed amount of myopic astigmatism at the time of surgery? Is it not further a self−defeating concept in that one must expect uncorrected vision generally of 20/30 or less for distance and near because of the induced refractive error? What then is the rationale for promoting this concept if, indeed, multifocal or bifocal intraocular lens technology proves itself clinically to afford the potential for 20/20 distance and near vision?

The concept of the multifocal intraocular lens has been approached with skepticism by some because bifocal contact lenses have had an overall poor performance. While no consistently successful precorneal bifocal lens has reached widespread use, these lenses are easily removed, unlike intraocular lenses. This simple fact probably stifled the development of the multifocal implant for some time.

However, it must be remembered that a precorneal lens is very dependent upon the external ocular environment: the precorneal tear film, lid positioning, and lens movement. The implant, on the other hand, is closer to the nodal point of the eye and, once implanted, tends to stay in a fixed position for the patient's lifetime. Unlike the contact lens, the implant is entirely independent of external factors. Therefore, the skeptic that argues against multifocal lens implants based upon the poor performance of precorneal bifocal contact lenses argues from a weak position.

Defining Terms — "Multifocal vs. Bifocal"

(Author's note: The following are excerpts from an article by Joseph Hoffman that appeared in the Feb. 1, 1990, edition of Ocular Surgery News.) The Food and Drug Administration (FDA) and its ophthalmic adviser, the Ophthalmic Devices Panel, are working to sort out the specific terminology that describes lenses of more than one focal length. "Multifocal" has always been a general term for such optical systems, but is "multifocal" misleading when referring to lenses that have only two focal lengths? The question arose at a meeting of the Ophthalmic Devices Panel early last year and the panel's consulting ophthalmologists answered "yes."

At the time of the meeting, only 3M, of St. Paul, Minn., and Iolab, of Claremont, Calif., had begun U.S. clinical studies of IOLs with more than one focal length, both lenses designed to produce two. However, other designs that employ aspheric and "holographic" optical properties to produce numerous focal lengths reportedly are nearing clinical study in the United States. . . .

"If you have a lens which is truly multifocal, it implies that you have several zones of clear focus beyond the standard bifocal, such as with the trifocal, which would be a multifocal lens," Claude Cowan, MD, another panel consultant said . . .

After more discussion a vote was taken among the ophthalmologist-consultants present, with five of the seven in favor of calling the two IOLs under discussion bifocals, none in favor of calling them multifocals and two abstaining. As a result of this vote . . . the FDA has been advising IOL manufacturers of the new restricted sense in which the term "multifocal" should be used.

The issue has apparently evaded final resolution, however. According to Suzanne Danielson, clinical regulatory supervisor at 3M Vision Care, the FDA has pursued the question of terminology in the course of recent communication with 3M. . . . Up to this point, Danielson said, 3M has simply adopted terminology for lens optics established by the American National Standards Institute (ANSI). In a document entitled "Prescription Ophthalmic Lenses—

Recommendations," approved by ANSI June 24, 1987, the following definitions are given:

"3.21.2 **Bifocal**, A lens designed to provide correction for two viewing ranges. . . . 3.21.12 **Multifocal,** A lens designed to provide correction for two or more viewing ranges."

"If the intraocular lens standards are going to be different than the spectacle standards, that seems like an issue for the FDA to work out with ANSI," Danielson said.

Former Ophthalmic Devices Panel member, Jack T. Holladay, M.D., stated in a recent interview that the optical design of bifocal and multifocal IOLs might end up being too academic a question for most clinicians to worry about. "We've never gotten to the point before that the specific optics [of an IOL] made that much difference to the patient or the surgeon," Holladay said, "and we're probably going to find that those differences in names of the lenses— whether bifocal or multifocal—are going to be irrelevant. My guess is that one of the bifocals and one of the multifocals are going to turn out to be the best in terms of performance. But it is not going to be because it's a bifocal or a multifocal, it's going to be the overall optics." Holladay characterized the narrowing of the terminology as an inconsistency. "[The FDA] is redefining "multifocal" to mean something that it never has in the past—three or more focal lengths rather than two or more," he said. . . .

The phenomenon of "pseudoaccommodation," in which researchers have found that IOLs designed to be bifocal appear in some cases to provide a range of clear vision between the two focal lengths, also muddies the issue. . . .

FDA sources explained that this is intended to clarify IOL design to ophthalmologists, and only clinical data will reveal the relative superiority of any given type of bifocal or multifocal IOL.

In researching ophthalmic terminology, senior OSN staff writer, Joseph Hoffman, consulted many reference sources, all of which confirmed that bifocal was a specific kind of multifocal lens system. Donald R. Sanders, M.D., PhD, Chief Medical Editor of Ocular Surgery News, probably stated it best in his note: ". . . unless and until the FDA requires IOL manufacturers to apply the more restricted definition, we will be free to use the term 'multifocal' in referring to lenses designed, as the ANSI statement says, 'to provide correction for two or more viewing ranges.' This is one case where usage is grammatically correct, or at least not grammatically incorrect."

Two very dissimilar lens designs have completed initial U.S. CORE studies and passed into expanded CORE status as of this writing.

Iolab Nuvue Bifocal Lens

J. McHenry Nielson, M.D., Associate Professor of Ophthalmology, University of South Florida, working at that time with Precision-Cosmet Corporation, developed a two–zone optical bifocal intraocular lens. The original lens was a three-piece Prolene™ looped plano convex lens with a UV-absorbing 7−mm PMMA optic. The lens functioned as a bifocal because it has two distinct optical elements. The base optical element provided distance correction and comprised the majority of the lens's optical surface. The central 2-mm portion of the optic, however, provided for near vision by incorporating an additional 4.00−diopter (in aqueous) correction. A lens of this original design was first implanted in Bromsgrove, England, by John Pearce in June 1986.

Iolab acquired Precision-Cosmet in December 1986, and the bifocal project was recognized as having great potential significance. During 1987, Mr. Pearce continued to evaluate the lens clinically, implanting it in more than 50 patients. In mid 1987, Richard Keates, M.D., Professor and Chairman, Department of Ophthalmology, University of California, Irvine, implanted the first bifocal implant of this design in Columbus, Ohio. In cooperation with Dr. Nielson, Dr. Keates assisted in modifying the original design to include a one-piece biconvex lens. This PMMA UV−absorbing product has a 7-mm optic with modified C-loops and is included in the expanded CORE studies in progress as of the time of this writing. As of April 1990, more than 1,000 Iolab bifocal lenses had been implanted worldwide.

Clinical results to date have generally been very favorable. The best corrected distance visual acuity at four to six months is shown in Table 21−1 and best corrected near visual acuity for the same time period in Table 21−2. Tables 21−3 and 21−4 show the best uncorrected distance and near vision in the nest case series, which excludes preoperative pathology and macular degeneration. Tables 21−4 and 21−5 show the best corrected distance and near results in this same subgroup of patients.

I have had experience with the Nuvue bifocal lens in 32 eyes. Patient acceptance and satisfaction with this lens has generally been good. I personally believe that patients who would be good candidates for a monovision−type situation function extremely well with this lens. I would avoid its use in patients with small pupils (smaller than 2.5 mm) or patients requiring miotic therapy. If the pupil is miotic, it is possible that good distance acuity would only be achieved by correcting the near add in distance spectacles. That is, if the eye saw clearly at distance at plano with a moderately dilated pupil and J1 at near without correction, one might expect to find a −2.75 refraction, which would achieve good distance vision. If the pupil was 2 mm or smaller, the patient might prefer this refraction for distance viewing rather than plano. Very careful patient selection and attention to pupil size should obviate the likelihood of this occurrence.

Patients receiving the Iolab bifocal in the FDA's study are routinely evaluated in both eyes for glare testing and contrast sensitivity. Glare is evaluated with the use of the Mentor brightness acuity tester on levels I and II in each eye. Results to date indicate no appreciable difference between the bifocal eyes and the fellow eyes that have undergone monofocal implantation. Contrast sensitivity is tested using the Vistech 8000 developed by Arthur P. Ginsberg, PhD.

Table 21–1. NuVue Bifocal Clinical Study Best* Corrected Distance Visual Acuity

% of Patients with 20/40 or Better at 4-6 Months

US	94.4	(51/54)
Europe	97.6	(239/245)
Total	97.0	(290/299)

* Excludes patients with preoperative pathology or postoperative macular degeneration.

February 1990

Table 21–2. NuVue Bifocal Clinical Study Best* Corrected Near Visual Acuity

% of Patients with J2 or Better at 4-6 Months

US	83.7	(41/49)
Europe	95.0	(230/242)
Total	93.1	(271/291)

* Excludes patients with preoperative pathology or postoperative macular degeneration.

February 1990

Table 21-3. NuVue Bifocal Clinical Study Best Uncorrected Far Visual Acuity at 12-14 Months

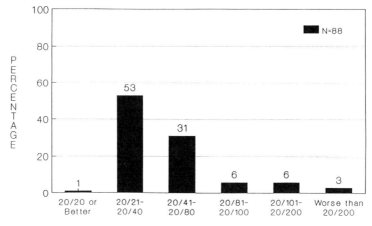

February 1990

Table 21-4. NuVue Bifocal Clinical Study Best Uncorrected Near Visual Acuity at 12-14 Months

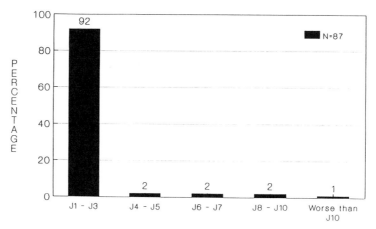

February 1990

Table 21–5. NuVue Bifocal Clinical Study Best Corrected Far Visual Acuity at 12–14 Months

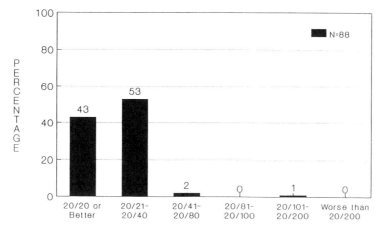

February 1990

Table 21–6. NuVue Bifocal Clinical Study Best Corrected Near Visual Acuity at 12–14 Months

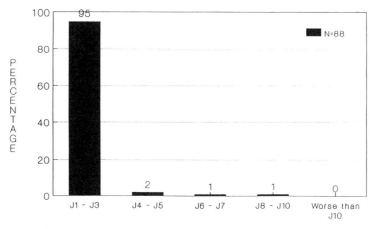

February 1990

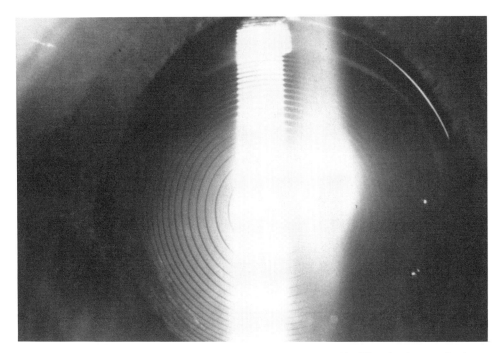

Figure 21–1. Clinical slit lamp photograph of the original design 3M diffractive intraocular lens as seen following pupillary dilation.

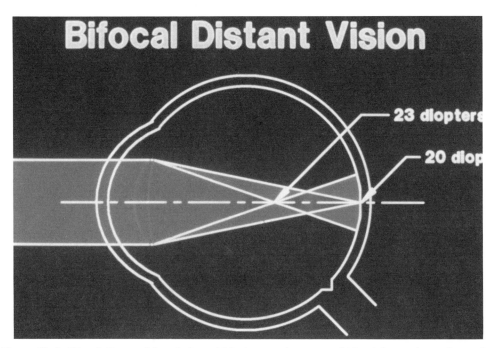

Figure 21–2. Bifocal distant vision during which the near image would be defocused anterior to the fovea and would not interfere with distance viewing.

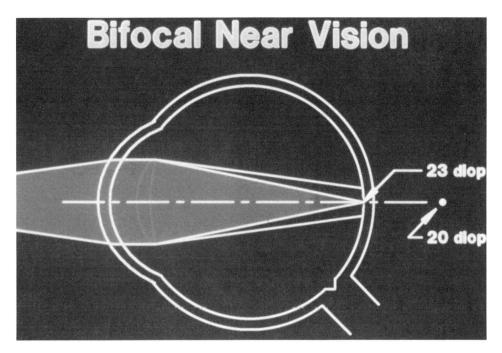

Figure 21–3. Bifocal near vision during which light from the near image is focused and light from a distant object would be defocused.

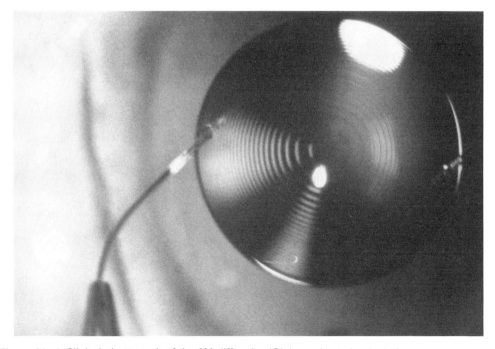

Figure 21–4. Clinical photograph of the 3M diffractive IOL just prior to implantation.

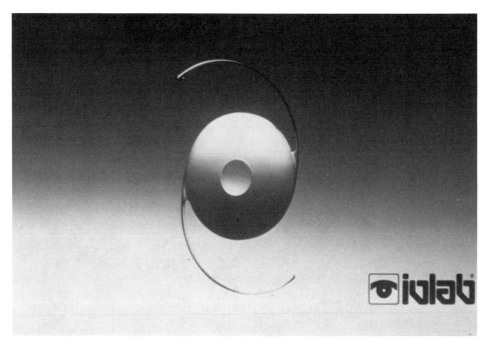

Figure 21–5. The original one piece plano convex IOLAB two zone bifocal design.

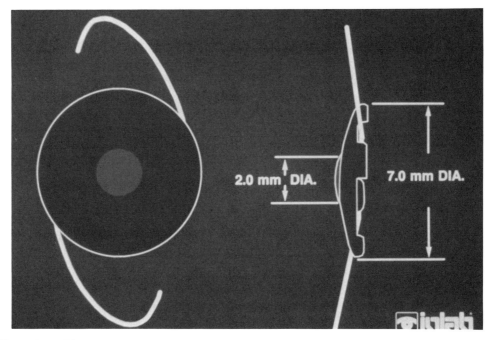

Figure 21–6. Diagrammatic representation of this lens showing the 2 mm central near zone and the 7 mm optic.

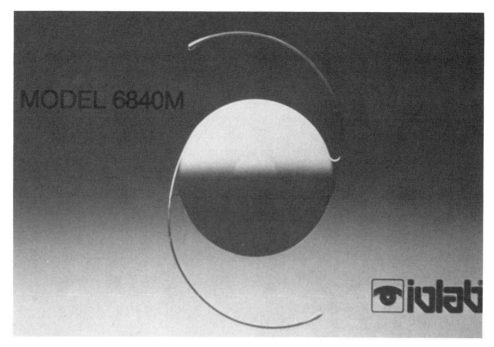

Figure 21–7. The later one piece biconvex two zone bifocal design.

Figure 21–8. A clinical photograph of this lens immediately prior to surgical implantation.

Patients are tested in both their bifocal eye and their fellow eye for both day and night conditions with and without glare.

3M Diffractive Multifocal IOL

Credit for the development of this unique design is given to William B. Isaacson, PhD, at 3M and a team of optical experts he assembled, including John Futhey, Mike Simpson, PhD, and Bill McAda. The initial design used a standard 6-mm meniscus PMMA UV–absorbing optic. Using a process of precise microreplication, a series of concentric diffraction zone plates or annular rings are superimposed upon the posterior surface of the optic. The number of concentric rings or annular areas as they appear determine the power of the near add. That is, the more rings or annular areas, the higher the power for near vision. The step height of each ring measures less than 2 microns, or about a third of the diameter of a red blood cell. Perhaps this is the reason that deposits have not been described as occurring within the annular rings.

The multifocal implant uses both refractive and diffractive principles. Refraction may be defined as the bending of light rays at any smooth optical surface. This occurs at the anterior surface of the 3M diffractive lens, as with any other standard monofocal implant. Diffraction, which occurs at the posterior IOL surface, is less familiar and deserves further explanation. Diffraction may be defined as the spreading of light or the spreading of a portion of a wave front as it encounters small surface discontinuities. The light waves that are in phase with one another will interfere constructively, resulting in high intensity at a given point. The waves that are out of phase interfere destructively and result in such low intensity as to effectively have been canceled out.

In other words, the foci are created by superimposing the microstructured surface or diffraction zone plate upon the posterior surface of the optic at such a precise spatial arrangement so as to create constructive interference of light waves that focus at two predetermined points dependent upon the desired lens power. Light passing through each zone contributes to both foci. This creates a balanced intensity lens that delivers 41% of the available light to a near focal point and 41% to a distance focal point while the remaining 18% of light is present but is focused at higher orders of diffraction.

Light energy passing through the lens essentially unaltered falls into the "zeroth" order, which is truly no diffraction at all, and is altered primarily by the implant's refractive anterior surface. Light in the first order of diffraction will be focused at a second point, which constitutes the near power for the lens.

It is critical to understand that this diffractive lens is not similar to a Fresnel optic, which is a refractive device. Light in a Fresnel lens passes to one or the other focus, but not both. A Fresnel optic is not considered to be a viable design for a multifocal intraocular lens.

The 3M diffractive multifocal lens was originally implanted in the U.S. by Dr. John Sheets, M.D., in Odessa, Texas, and soon thereafter by six other U.S.

investigators. A 6–mm meniscus PMMA optic with closed loop polypropylene haptics was implanted in the first 50 U.S. cases by the original group of seven U.S. investigators. The first lens was implanted on Nov. 13, 1987, and most of the 50 original patients have completed their second postoperative year.

As of mid 1990, over 25,000 diffractive multifocal lenses have been implanted worldwide, including in two major clinical studies. The FDA's expanded CORE study has completed enrollment and encompasses approximately 700 patients under a strict FDA clinical protocol that requires one year of followup in all cases. A major European multicenter study has also been completed that involves 16 surgeons from 15 European countries and about 350 patients. The results of both the European study and the FDA's study appear to agree closely and complement each other.

Distance acuity in both studies meet or exceed the FDA grid rate for monofocal lenses. In comparison to an FDA grid rate of 94%, data show that 97% of patients without pathology see 20/40 or better. In addition, functional near acuities (J1 to J3) are achieved in more than 85% of these patients without any add (Table 21–7).

Glare testing using the Mentor brightness acuity tester has failed to show any significant difference in a series comparing the multifocal and fellow monofocal eyes. One of the primary concerns with multifocal intraocular lenses of either diffractive or refractive design is whether the quality of distance vision is truly comparable to that obtained with standard monofocal intraocular lenses. The key issue regards reduced image contrast, which results from the splitting of light to distance and near focal points.

Contrast sensitivity with the 3M lens has been evaluated by four different methods. In Germany, Professor Dr. K.W. Jacobi used the Ocutrast to examine mesopic visual acuity in a group of 43 multifocal patients and found no difference in contrast sensitivity compared to a control group of patients with monofocal IOLs. His results were presented at the meeting of the American Academy of Ophthalmology in 1989 in New Orleans. At the same time, I reported on the results of a series of 26 patients with a diffractive multifocal lens in one eye and a monofocal lens in their fellow eye. The best case subgroup of these patients showed no statistically significant difference in their performance on the Pelley-Robson contrast sensitivity chart.

In the Journal of Cataract and Refractive Surgery in January 1990, Tom Olsen, M.D., and Leif Corydon, M.D., reported on 13 pseudophakic eyes in ten patients implanted with the diffractive multifocal IOL and compared them with an age–matched control group with conventional monofocal IOLs. Using the Vistech, Inc., test charts, they found no significant difference in contrast sensitivity for distance vision and suggested the finding indicates an uncompromised distance focus for the multifocal IOL. They did find an overall decrease in contrast sensitivity of 0.19 log units in the multifocal group when these patients were tested without near add and compared with the controls with the near addition.

We continue to study paired multifocal and monofocal eyes with various methods, including Regan charts, Pelley-Robson charts at ten feet, and the Vistech System in the U.S. and the Ocutrast System in Europe. While the theoretical loss of contrast at each focal point will be demonstrable on some test methods, there is still doubt whether this bears clinical relevance to the patients' lifestyles, well being, and general acceptance of the lens.

A question has been raised as to whether the diffractive zone plates on the posterior implant surface could create any problems with retinal visualization or laser treatment either of the retina or posterior capsule. I have personally had experience with each of these situations and can verify that retinal examination and photocoagulation is easily accomplished through the diffractive lens. Additionally, there has been no difficulty in performing YAG laser posterior capsulotomy in the diffractive implant population as compared with monofocal eyes. Furthermore, we have not noted deposition of particulate matter in circumferential patterns on the posterior surface of the diffractive lens. Prior to initial implantations, this concern had been voiced; however, when one considers that the step height of the diffractive rings is about ⅓ the diameter of a red blood cell, it is easy to realize how they would not lend themselves to deposition of particulate matter.

Other Lens Designs

While the two–zone Iolab lens and the 3M diffractive lens are the only ones to have entered expanded clinical studies in the U.S. to date, additional lenses are attempting to gain IDEs for study, and one lens has begun initial CORE evaluation.

In cooperation with Irvine M. Kalb, M.D., Pharmacia has developed a multifocal lens with a central distance zone surrounded by a donut shaped optical near zone peripheral to which the remainder of the optic provides distance correction.

Lee Nordan, working with Wright Medical Inc. with silicone and Ioptex with PMMA, has defined an aspheric multifocal lens. Studies of the aspheric lens have begun in Europe, but not in the U.S.

Conclusion

I conclude this chapter with several points in which I now believe. (1) Multifocal implant technology is evolving, and we should expect to see one or more premarket approved designs available within the near future. (2) By definition, a reduced contrast image is created with all multifocal lenses. While this does not limit the potential resolution to less than 20/20, the 20/10 potential "crispness" may be missed by some patients. (3) We can expect the development of small–incision multifocal lenses in the near future as phacoemulsification continues to become more widely accepted as the procedure of choice for cataract extraction.

While we may hear many theoretical concerns voiced as to possible limitations of multifocal or bifocal lenses, in the final analysis, patient acceptance will determine the success or failure of these devices.

22

IOL Insertion and Wound Closure

James A. Davison, MD

After capsular vacuuming, the anterior chamber is filled with enough viscoelastic to create a minimal posterior capsule concave posture (Figure 22–1). The incision is extended to 5.5 mm with a super blade. A smaller extension to 4 mm may be necessary if a foldable IOL is to be used, and the incision will have to be a little larger if a 7 mm optic is to be employed. I perform a last–minute check of the IOL power calculation sheet, which is written in my own hand, and match the desired IOL power to that on the IOL container and the eye and name to the patient (Figure 22–2).

I usually use a one–piece all PMMA biconvex 6 mm UV–filtering optic in a short C-loop 12 mm overall length with a six–degree posterior angulation (Figure 22–3). It is very soft and is easily placed within the capsular bag with a Lester hook (Figures 22–4, 22–5). No manipulation of the superior haptic is required for insertion.

This configuration centers extremely will in large or small eyes. At surgery, there is no distortion of the anterior capsular remnant, capsular equatorial zone, or posterior capsule. There is no tendency to generate ciliary body irritation secondary to capsular bag overstretch, which can be seen with capsular bag implantation of longer, stiffer lenses better suited for ciliary sulcus implantation (Figure 22–6). Over the long term, a 12 mm soft C design seems to help balance the force of capsular contracture with the capsular bag expansion tendency created by peripheral zonular traction. This results in stable long–term optic centration. For very large eyes or moderate to larger eyes in younger patients, I generally use a 7 mm IOL optic with a very soft 13 mm overall haptic length.

A C-loop 12 mm haptic configuration makes IOLs easy to implant when operative conditions are good. More importantly though, these lenses are safer to implant than longer, stiffer designs when the going gets tough. The unwelcome but not infrequently encountered situations of the short eye, tight eye,

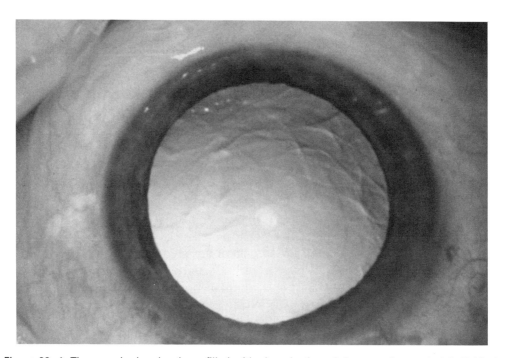

Figure 22–1. The capsular bag has been filled with viscoelastic and the wound extended. Individual ribbons of viscoelastic can be seen. The posterior capsule is just concave, and the bag has not been overfilled.

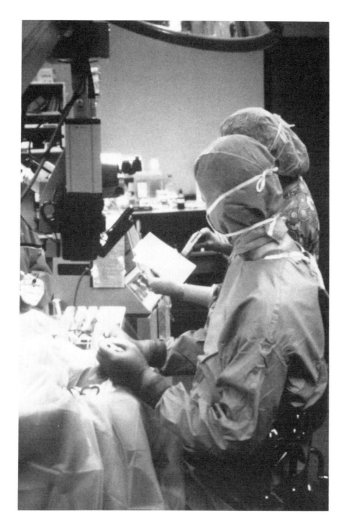

Figure 22–2. A final last–minute check is made of the original calculations sheet, name, and eye with the IOL to be implanted.

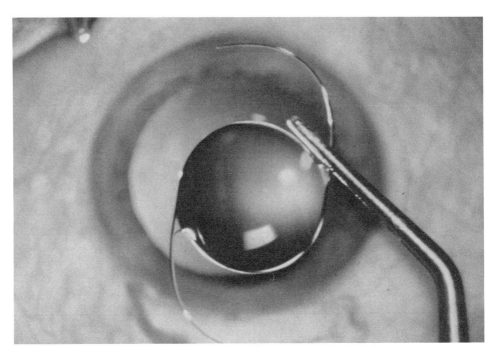

Figure 22–3. My IOL preference is a one–piece all PMMA, biconvex, UV–filtered, 6–mm optic in a 12–mm overall, six-degree optic posterior angle, soft C-loop design.

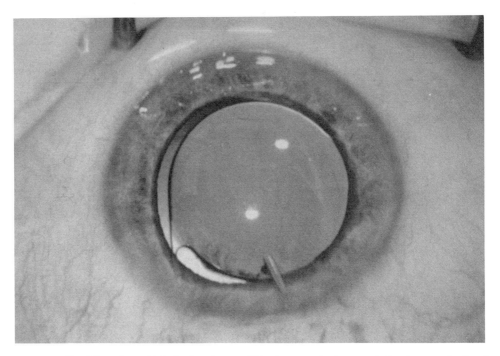

Figure 22–4. The IOL optic and inferior haptic are within the capsular bag ready for the superior haptic to be placed.

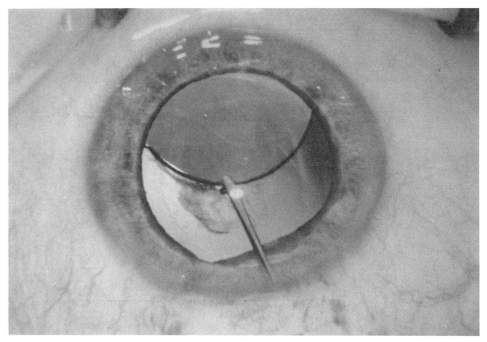

Figure 22–5. The IOL is inserted, making sure that the inferior loop is underneath the anterior capsule. The ease of inferior haptic compression permits the Lester hook to dial in this soft lens without having to manipulate the superior haptic.

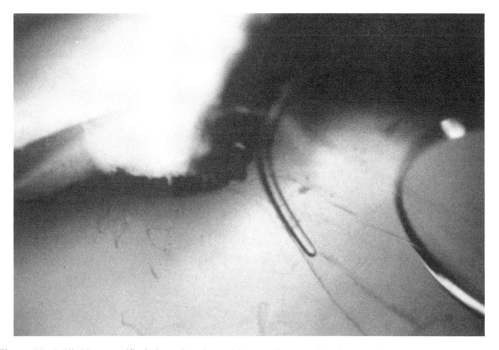

Figure 22–6. Highly magnified view showing a 12–mm diameter haptic contained within the capsular bag but not pushed against the ciliary processes by capsular bag overstretch. (Photograph taken at Laboratory For Intraocular Lens Research, Charleston, S.C.).

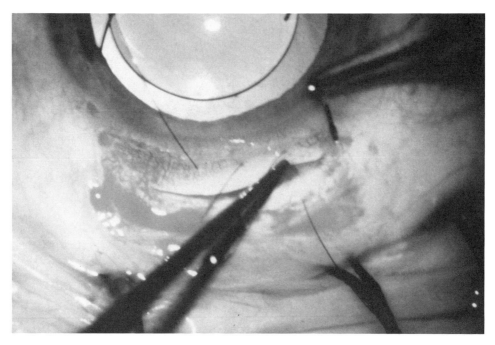

Figure 22–7. Suturing is accomplished with shoelace 10-0 nylon. A total of seven bites are usually taken. A surface knot is used to avoid unnecessary wicking with a buried knot.

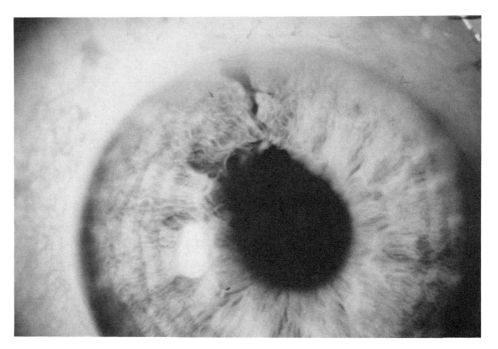

Figure 22–8. Sector iridectomy has been repaired with two 10-0 Prolene sutures on straight needles as described by Worst.

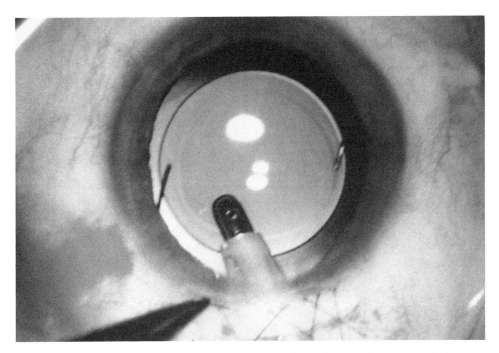

Figure 22-9. The BSS infusion bottle is lowered to 40 cm above the eye. The 0.3-mm tip is used on a maximum fast setting to remove as much viscoelastic as possible.

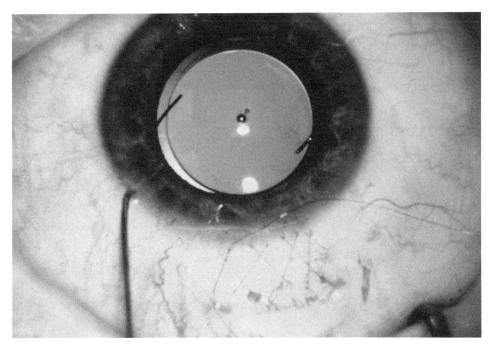

Figure 22-10. The sutures have been brought up and secured, and a triple throw knot is created. The anterior chamber is inflated with BSS through the stab incision with a 30-gauge cannula.

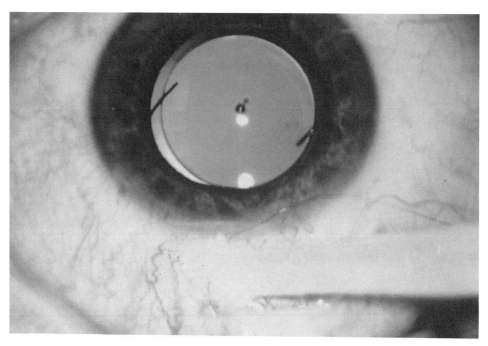

Figure 22-11. The wound is tapped and checked for leakage and the eye checked for firmness. It should have normal pressure of about 20 mm—25 mm of mercury. For practical purposes, this is largely done by feel.

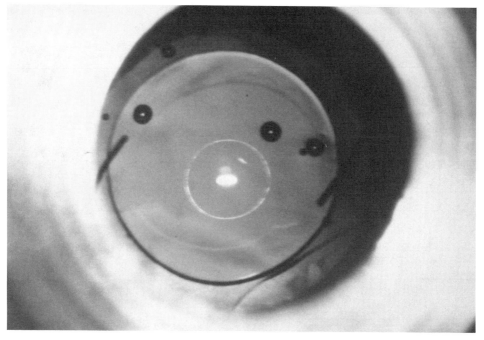

Figure 22-12. The Maloney keratoscope is used and a horizontal ovoid reflection seen, indicating perhaps three to four diopters of induced with-the-rule astigmatism.

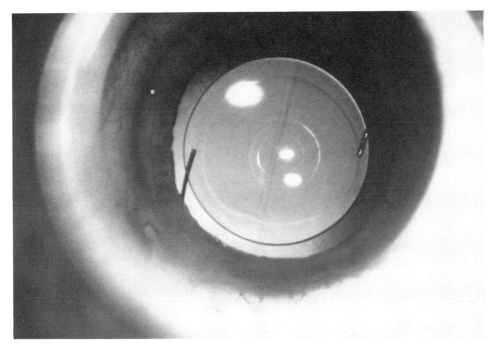

Figure 22–13. After suture adjustment, reinflation of the eye, and rechecking of the pressure with a sponge, the Maloney keratoscoptic reflection image is more circular, indicating the cornea is almost spherical.

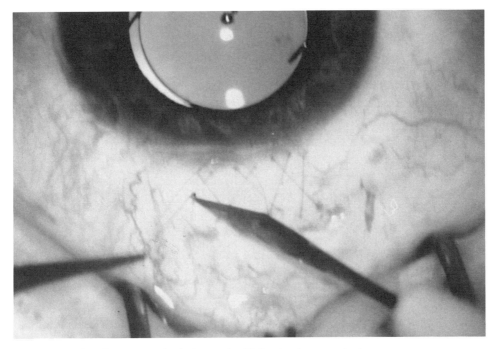

Figure 22–14. After three more single knots, the 10-0 nylon is cut at the knot.

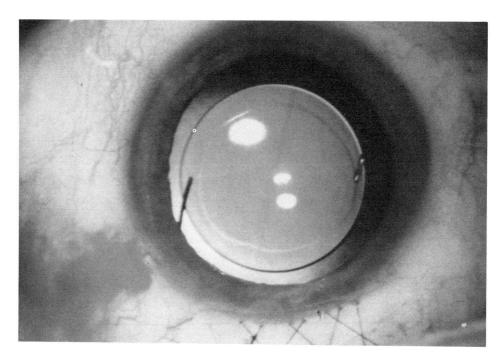

Figure 22–15. A three-piece IOL is in good position, and the wound is closed.

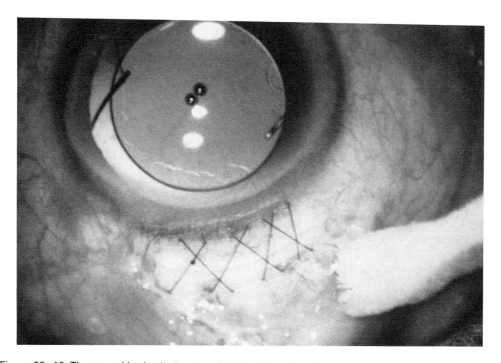

Figure 22–16. The wound is checked and a slight leak found at the extreme of the right-hand wound.

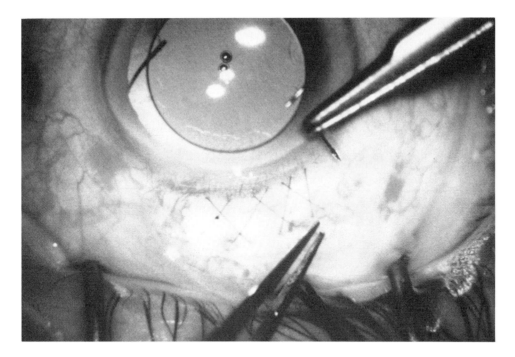

Figure 22–17. An extra suture is placed.

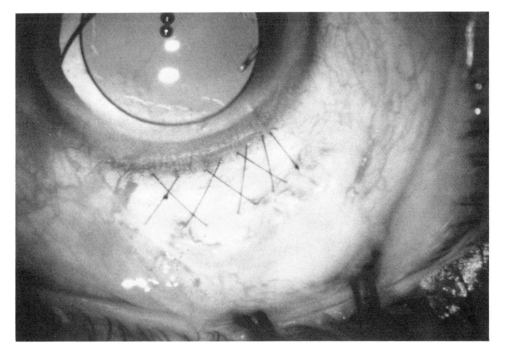

Figure 22–18. The wound is now improved and watertight. (The irregularity in the wound construction can be seen in Figure 22–   at the entry of the 5520 blade).

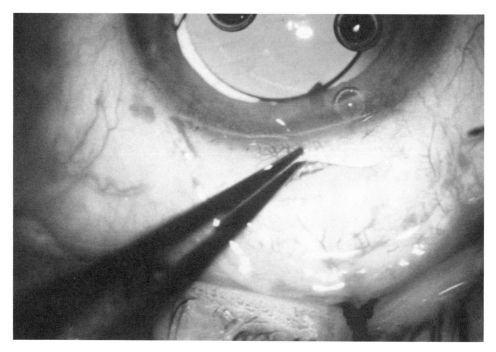

Figure 22–19. Some anterior chamber bleeding can be seen with the scleral flap incision as it is inspected prior to closing.

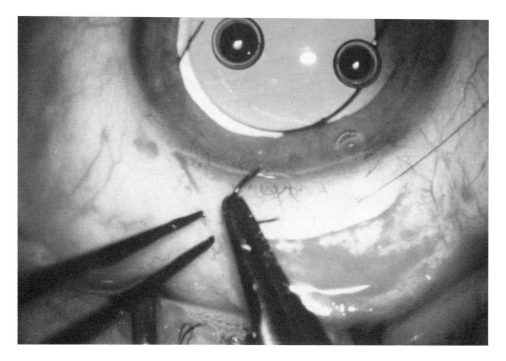

Figure 22–20. A central transverse 10-0 nylon suture can be placed without observation of the breadth or depth of deep wound engagement.

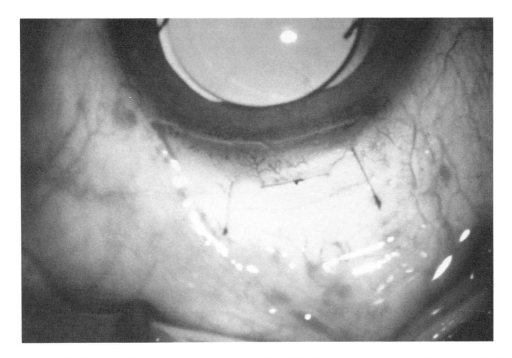

Figure 22–21. One radial suture is placed on each side of the transverse suture, creating a watertight closure with no induced keratometric astigmatism.

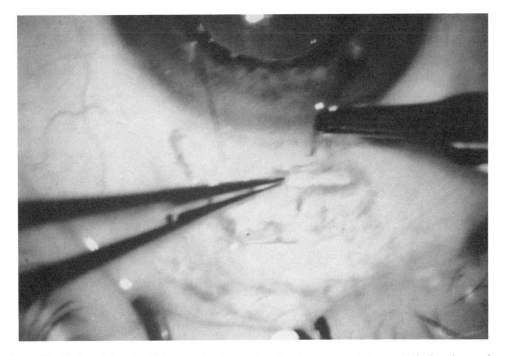

Figure 22–22. A peripheral radial suture has been placed at the extreme right- and left-hand wounds. The initiating bite in the anterior flap is taken in preparation for the transverse suture.

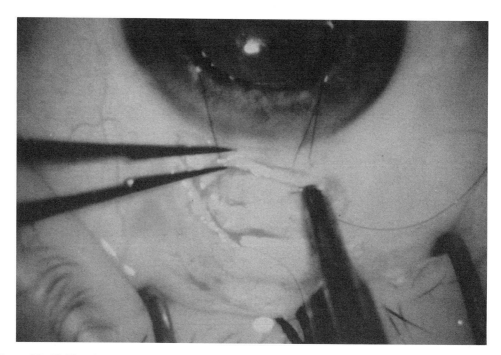

Figure 22–23. The deeper bite of the transverse closure is accomplished under direct visualization. This aids in engaging adequate tissue but not driving too deep or at an off angle.

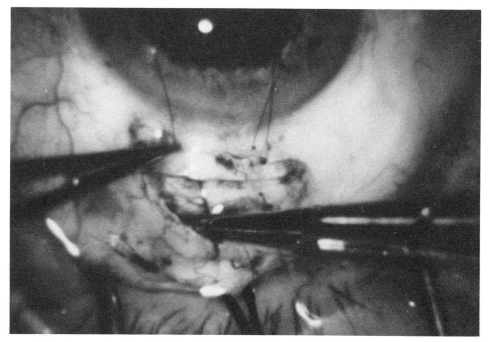

Figure 22–24. The needle has been regrasped backhand and the overlying scleral tissue flap draped over it at the appropriate location.

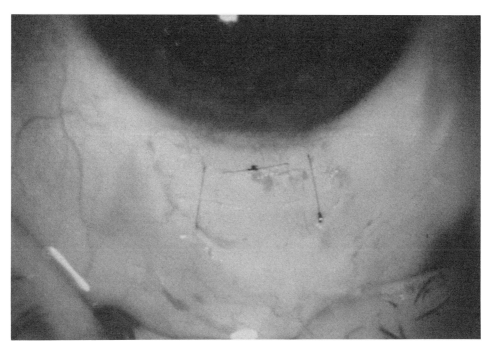

Figure 22–25. The final closure has been accomplished. The transverse suture is not placed perfectly, but a watertight closure has been completed.

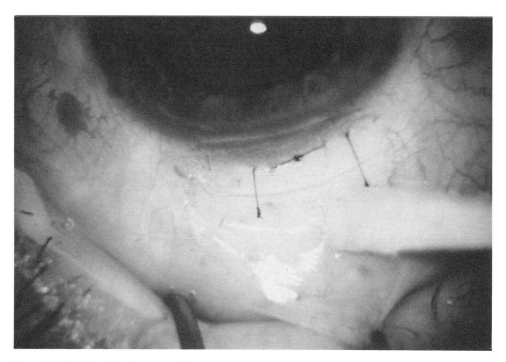

Figure 22–26. A leak in the wound has been exposed by applying gentle pressure on the posterior wound edge.

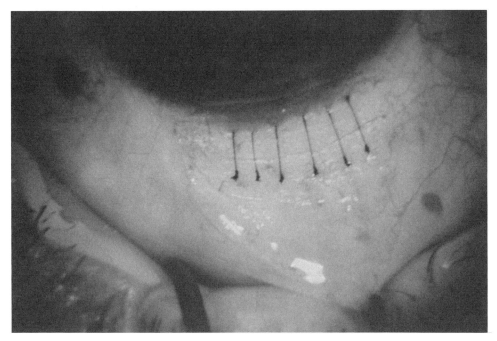

Figure 22–27. Multiple radial sutures can be used to better close an incision that appears to leak with the hybrid radial-transverse closure. Six sutures have been used here, but as few as four may suffice.

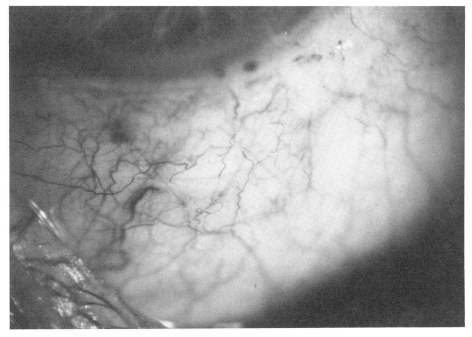

Figure 22–28. Three millimeters back, the double X pattern approximates sclera to sclera peripheral to the limits of the anterior chamber and thus helps avoid leakage. Each bite is at least two-thirds depth (the depth of the initial groove and flap dissection) which permits good approximation without excessive compression.

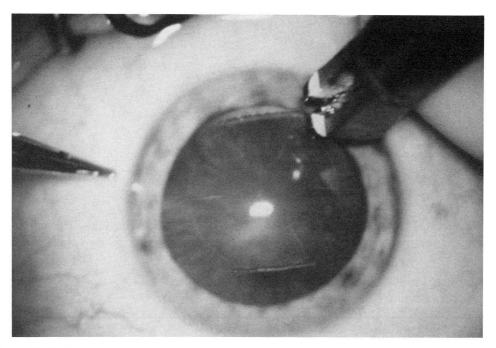

Figure 22–29. Horizontal TAK incisions are created within the cornea and, at 3 mm from the visual axis, are relatively close to it, which greatly influences astigmatism.

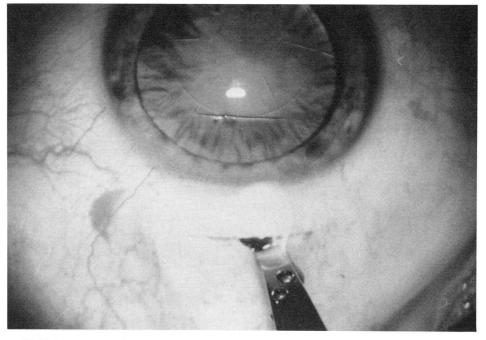

Figure 22–30. In contrast to the TAK incisions, the beveled horizontal IOL wound starts 3.0 mm posterior to the limbus and has little influence on keratometric astigmatism.

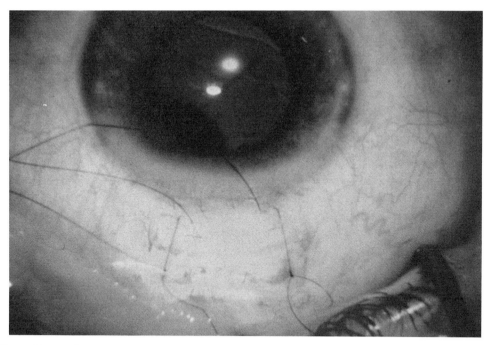

Figure 22–31. Some bleeding may be encountered with posterior wounds and their closures, but irrigation will wash all of the blood from the anterior chamber.

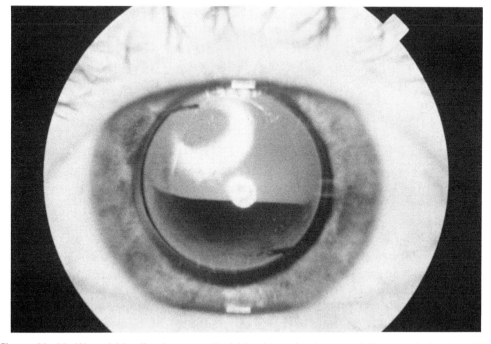

Figure 22–32. Wound bleeding has permitted blood to migrate around the capsulorhexis and be trapped behind the IOL and the posterior capsule. The vision in this case is still 20/20, but the blood was slow to resolve.

small pupil, small capsulorhexis, imperfect zonular support, very steep brow, or merely the restless patient are all more easily managed with a softer, shorter haptic IOL configuration.

Suturing with 10-0 monofilament nylon can be accomplished with a conventional seven–bite running closure and a surface triple throw knot (Figure 22–7). Sector iridectomies are repaired using a straight needle and 10-0 Prolene in a fashion described by Jan Worst (Figure 22–8). Viscoelastic substances are then removed with the 0.3 mm I/A tip and the BSS bottle still lowered to 40 cm above the eye (Figure 22–9).

If a traditional running suture pattern is employed, final suture tightness should be obtained with a keratometric control. The sutures are first tightened empirically with the eye inflated fairly firmly with a 30–gauge cannula through the sideport incision (Figure 22–10). A sponge is used to tamp down the wound while expressing small amounts of BSS through the wound edge to check alignment and tightness (Figure 22–11).

When the eye is at a "normal" firm pressure, a Maloney keratoscope is used to check the keratoscopic reflex (Figure 22–12). In the case of Figure 22–12, the sutures have been pulled too tight, and it appears that about four to five diopters of with-the-rule keratoscopic astigmatism have been induced. The suture is then adjusted very slightly so that it is ever so slightly more lose. The steps of inflation, tamping, and keratoscopic check are then repeated, yielding a more circular image (Figure 22–13). Any amount of against-the-rule astigmatism (vertically ovoid reflex) should not be tolerated.

A well centered IOL and a secure wound should be the results of these efforts (Figures 22–14, 22–15). Myopes with thin sclera, imperfect wound design, or too tight a phaco incision with a long phaco time may occasionally yield a wound that cannot be closed well without an extra suture. Inspect for leaks after the final closure; if any are detected, place an extra suture (Figures 22–16, 22–17, and 22–18).

Jim Gills (personal communication, July 1989) suggested that John Shepherd's transverse suture technique may be used centrally in combination with a radial suture on each side to create a secure wound with very minimal amounts of induced with-the-rule astigmatism. This combination of suturing techniques was a terrific idea and it works very well. I now use it, and it has become one of my standard closures (Figures 22–19, 22–20, and 22–21).

I first place a radial 10-0 nylon on each side of the wound. I drive the needle through the superficial flap in radial fashion of the right side (Figure 22–22). I go all the way through the flap and regrasp the needle. I then drive the needle through the deeper scleral bed in transverse fashion (Figure 22–23). I make sure that it emerges in the correct spot and then drape the superficial flap over it (Figure 22–24, 22–25). I grasp the sharp portion of the needle and complete the bite. If done in this fashion, direct visualization of the deeper flap needle pass will give a more accurate consistent bite. This has worked very well for me and almost always produces one diopter or less of induced keratometric change on the first day of surgery. Followup examinations two weeks later show about one diopter of induced with-the-rule astigmatism, and at two months or earlier there is a decay back to the preoperative keratometry readings.

I have had to replace the transverse suture with interrupted ones if there is any hint of leaking at the time of closure (Figures 22–26, 22–27). This has led me to favor using two "X" pattern mattress sutures with both knots pulled back to the posterior bites (Figure 22–28). When a 5.5 mm wound is placed 3.0 mm from the limbus, only about 1.0 diopters of keratometric astigmatism is induced on the first postoperative day. The minimal compression effect decays rapidly and the wound is stable long-term. The two X's represent a good closure compromise between convenience and effectiveness. It is convenient for two reasons. Only two knots are required and it rarely needs to be replaced or adjusted because of leakage seen at surgery. It is effective because it uses less 10-0 nylon in each length than a seven bite running suture and thus avoids almost as well as short interrupted sutures the wound compressive effect generated by the increased total elasticity (as measured by total change in length) of the longer nylon segment needed in a running closure.

With the wound 2.5 to 3.0 mm back from the limbus, I seem to get just as good of an astigmatic result with four interrupted sutures as I do with the hybrid radial-transverse closure. The location and perfection of wound construction are far more important than its method of closure both short– and long–term. A well constructed wound will yield no net astigmatism effect, and this stability allows it to be combined with transverse astigmatic keratotomy to substantially influence preoperative astigmatism (Figures 22–29, 22–30).

The only disadvantage to a posterior wound is a greater tendency for bleeding. Prompt and thorough irrigation will usually take care of this problem. Rarely, postoperative hyphemas and temporary disappointment in immediate postoperative vision can be encountered (Figures 22–31, 22–32).

23

The Infinity Suture

I. Howard Fine, MD

The use of a horizontal mattress suture as first introduced by Dr. John Shepherd of Las Vegas, Nevada, has improved upon an already dramatic decrease in surgically induced astigmatism following small-incision intraocular lens implantation.[1] The possibility that a similar principle could be used for conventional 6.5 mm phacoemulsification incisions has been intriguing. Over the past six months I have used a variation of Dr. Shepherd's suture, which appears to show tremendous promise.

The technique involves the use of a scleral tunnel incision. Following phacoemulsification and intraocular lens implantation, the incision is closed by suturing the roof of the scleral tunnel to the floor of the scleral tunnel with a suture that, in cross section, resembles the mathematical symbol for infinity. One loop is placed from the right to the left with a bite that is approximately 40% of the tunnel width (Figure 23–1). The suture is then brought to the opposite end of the incision, and a second loop is placed going toward the first loop, again with a bite incorporating approximately 40% of the tunnel width (Figure 23–2). The two ends of the suture can be pulled, closing the left side of the incision while removing residual viscoelastic with the I/A handpiece (Figure 23–3).

Alternatively, the left half of the tunnel can be held closed by an instrument such as *Colibri* forceps for the viscoelastic removal. Following that, the suture is pulled up tightly and tied without keratometric control, and the suture ends are cut flush with the knot (Figure 23–4). The conjunctival flap is then draped back over the incision. I now use nylon. The great advantage of this type of closure is that no suture crosses the lips of the incision radially and, therefore, no force is exerted at the limbus in such a manner as to alter corneal curvature.

Alternatively, the compressive forces of the suture are tangential to the cornea and exhibit little or no effect on corneal curvature. The fact that the lips of the incision gape considerably after the suture is tied may appear

[1]Shepherd, John R., "Induced Astigmatism in Small Incision Cataract Surgery," Journal of Cataract and Refractive Surgery, 1989, 15:1, 85-88.

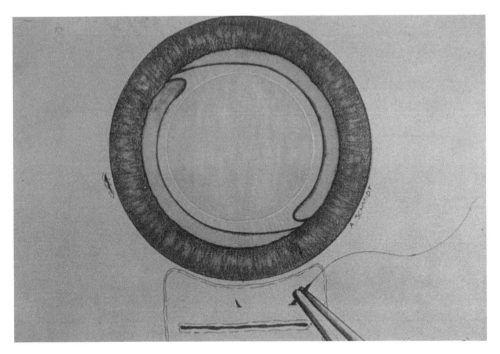

Figure 23–1. A loop of suture is passed from right to left with a bite that is approximately 40% of the tunnel width.

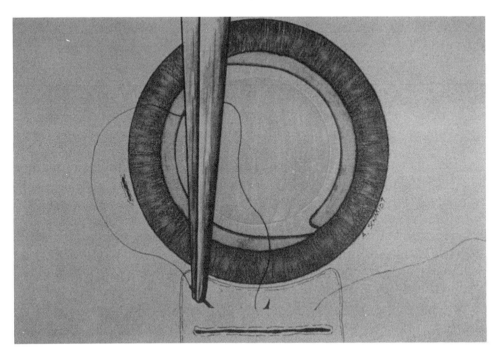

Figure 23–2. The suture is brought to the opposite end of the incision, and a second loop is placed going in the other direction.

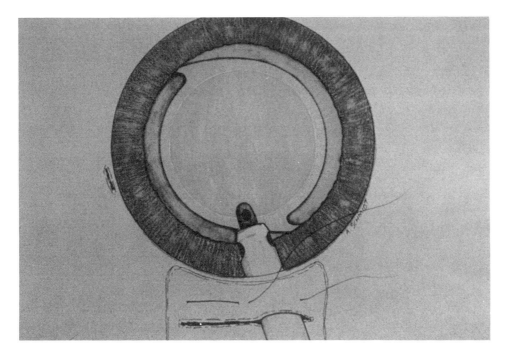

Figure 23–3. The two ends can be pulled, closing the left side of the incision while removing the residual viscoelastic.

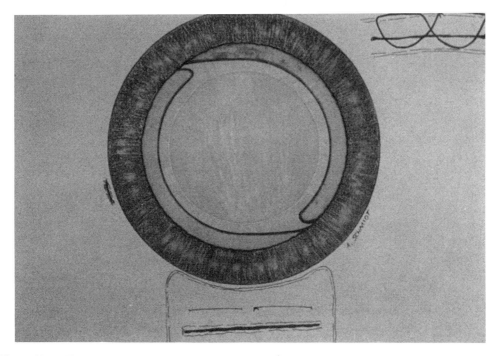

Figure 23–4. The suture is tied tightly and the ends cut flush with the knot.

intimidating to some surgeons. One should bear in mind, however, that following the placement of a groove and the dissection of a scleral tunnel up to the point of entering the anterior chamber, there is considerable gaping of the lips of the wound, and the K-readings have not changed at all.

The preliminary results for this suture have been excellent. I have used it in approximately 100 patients. I have not experienced any wound leaks, filtering blebs, or anterior chamber depth shallowing. Within a day or two of the time of surgery, the postoperative cylinders are within a half a diopter of the preoperative cylinders. There has been very little regression over the first two months postoperatively.

What remains to be seen, through controlled studies, is what the regression of this type of incision closure will be over a period of several more months to several years. However, the initial trial with this suture is very promising.

Section X
OTHER TECHNIQUES

24

Vitrectomy Techniques

Paul S. Koch, MD

The whole concept of vitreous loss requires some consideration when it occurs during phacoemulsification, because the vitreous acts differently here than it does in nucleus expression procedures. The closed system of phaco-emulsification has certain advantages in limiting the damage caused by vitreous loss and in enhancing the surgeon's ability to perform an appropriate vitrec-tomy.

If vitreous loss occurs in a nucleus expression procedure, there are no natural forces to limit the flow of vitreous out of the eye. Vitreous loss is directly related to the size of the incision and to the liquidity of the vitreous gel. In phacoemulsification, on the other hand, the constant pressure in the anterior chamber helps to hold the vitreous back, and the small incision limits how much vitreous can extrude from the eye.

The pressure in the anterior chamber in a closed system can even prevent vitreous loss. If the posterior capsule is opened during the operation, but the vitreous face remains intact, the anterior chamber pressure can hold the vit-reous back and prevent vitreous loss. If the vitreous face is disturbed and vitreous is prepared to move forward, the pressure in the anterior chamber can hold it back, preventing loss, or at least limiting the vitreous movement to a mild prolapse into the anterior chamber (Figure 24–1).

True vitreous loss can occur if the posterior capsule open suddenly, or if it opens in such a way that the surgeon doesn't recognize it immediately, such as if it occurs behind a piece of nucleus so the surgeon cannot see it. The aspiration port of the phacoemulsification handpiece will remove fluid from the anterior chamber, and the fluid will be replaced partially by solution from the irrigation bottle and partially by vitreous. Further aspiration may result in vitreous being aspirated to the phaco tip, with further loss. In a sense, the amount of vitreous loss is directly related to the time between the opening of the posterior capsule and the moment when the surgeon recognizes it (Figure 24–2).

This discussion of vitreous loss and vitrectomy techniques will revolve around an analogy involving a Slinky toy. You may remember this toy; it's been around for several decades. It's a floppy spring that can be used in a variety of playful ways. It comes in a variety of sizes and colors just as the vitreous comes with many individual characteristics, but for now we will consider an "average" Slinky as representing an "average" vitreous.

One good example of how the vitreous acts like a Slinky occurs in the vitreous wick syndrome. A small, narrow band of vitreous is stretched to the incision. Tension is placed right on through the vitreous body, disturbing the retina, and cystoid macular edema develops. Severing the vitreous band near the incision causes the rest of it to retract quickly in the eye, like the retraction of a spring. The cystoid macular edema often will resolve.

Another example is a Weck Cell vitrectomy. With the Weck Cell, some vitreous can be pulled out of the eye where it is cut. The "lost" vitreous retracts quickly into the eye.

We will consider the study of vitreous dynamics by imagining it as a tiny Slinky. We will imagine that a Slinky is attached to the vitreous base and another attached to the vitreo-macular interface. If the Slinky is pulled just a little bit, the forces will be absorbed by the first few coils of the spring and not much will happen (Figure 24–3). If it is pulled a lot, the entire coil will stretch out and exert tensions at its base (Figure 24–4). In the case of our imaginary Slinkies, that would put tension on the vitreous base, leading to a detached retina, and on the vitreo-macular interface, leading to cystoid macular edema.

This leads us to the first principle of vitrectomy:

The First Principle of Vitrectomy:

Do Not Stretch the Slinky

The best way to avoid stretching the Slinky is to maintain an intact posterior capsule. If that is not possible, we have to maintain as much of the integrity of the vitreous as possible by recognizing the tear and stopping aspiration. This may prevent any vitreous loss. If vitreous moves into the anterior chamber, however, a vitrectomy will be necessary.

The vitrectomy should be performed so that the Slinky is not disturbed. Specifically, this means that the vitreous body must not be violated during the vitrectomy either by the infusion or the aspiration.

If the vitrectomy is performed with a coaxial infusion cannula slipped over the vitrectomy tip, a one-handed vitrectomy can be performed, but this disturbs the Slinky three ways (Figure 24–5).

Three Ways a Coaxial Infusion Vitrectomy Stretches the Slinky

1. Extension of the Posterior Capsular Tear

The force of the infusion is directed in the same direction as the tip. That means the infusion will be directed downward into the deep areas of the eye. As the tip passes down toward an opening in the posterior capsule, the infusion flow will strike the capsular flaps, forcing them apart. This separation extends the capsular tear and enlarges the opening. This permits more vitreous to prolapse forward and stretches the Slinky (Figure 24–6).

A small capsular tear limits the forward prolapse of vitreous because the rest of the posterior capsule serves as a buffer barrier. Enlarging the tear reduces the size of the buffer barrier and permits more vitreous prolapse. Traditionally, we have considered the posterior capsule as the only structural barrier to vitreous movement, but the anterior capsule is also a barrier. There is not a lot of barrier if the anterior capsule has a large capsulotomy, but a strong barrier exists with a small capsulorhexis. The small opening in the anterior capsule will not rip open as the vitrectomy tip is inserted, and the barrier's integrity is maintained.

Trying to get vitreous to prolapse through a 4.5–mm capsulorhexis is like trying to squeeze toothpaste through the top of the tube. The small, rigid opening restricts movement and limits vitreous prolapse. This limits how much you stretch the slinky and may turn out to be one of the most significant benefits of the capsulorhexis capsulotomy.

2. Hydration of the Vitreous

The infusion fluid hydrates the vitreous, increasing its volume and causing it to expand. The only direction the vitreous is able to expand is toward the anterior chamber through the opening in the posterior capsule. This forward motion stretches the Slinky (Figure 24–7).

3. Flushing the Vitreous

The force of the infusion acts like a hydraulic hose and pushes the vitreous around, shaking and wiggling it, forcing the Slinky every which way, and exerting traumas to the Slinky's bases, the vitreous base, and the vitreo-macular interface. The vitreous is moving even in low flow systems, creating micro-traumas (Figure 24–8).

The result of this movement is the vitreous is flushed out of the back of the eye toward the anterior chamber. This flush pulls the body of each Slinky toward the anterior chamber, increasing the amount of vitreous that needs to be removed. This is what happens when what looks like a small vitrectomy turns into a large one, as the anterior chamber is constantly refilled with new, previously untouched vitreous. Eventually, little additional vitreous moves forward because a lot of it has been removed by then (Figure 24–9).

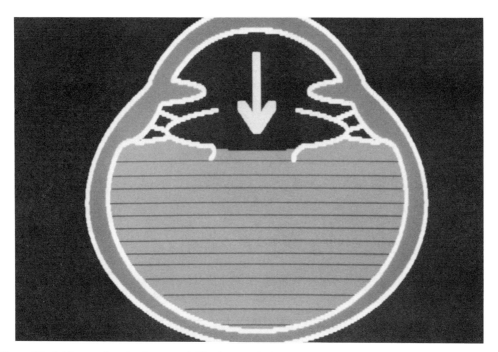

Figure 24–1. Vitreous loss in phacoemulsification can be limited because of the closed system. Pressure in the anterior chamber will press against the vitreous and hold it back.

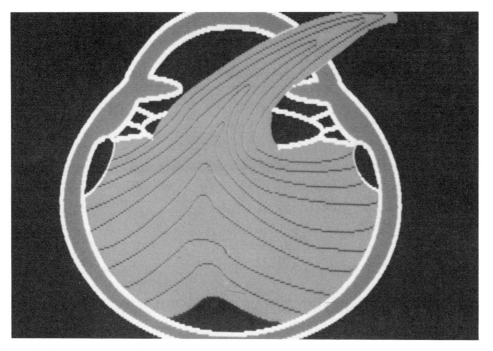

Figure 24–2. In procedures without a closed system, such as nucleus expression ECCE and ICCE, the amount of vitreous loss is not contained and can be significant, exerting tensions on the vitreomacular interface and the vitreous base.

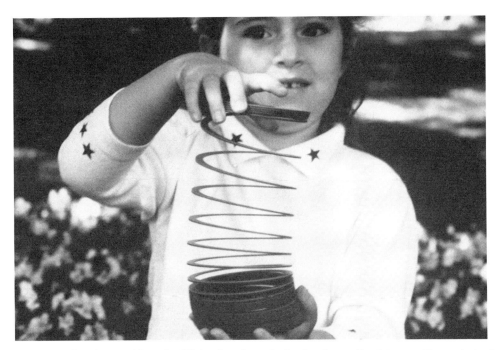

Figure 24–3. Gentle stretching of a Slinky disturbs only the first few coils of the toy. The main body is undisturbed.

Figure 24–4. If there is a lot of tension on the Slinky, all of the coils are distrubed and the tensions are reflected right down to its base.

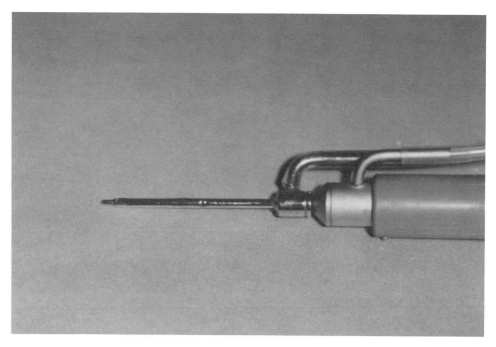

Figure 24–5. Vitrectomy tip with a coaxial cannula in place. Don't do this.

Figure 24–6. Coaxial infusion rips open the posterior capsule, permitting more vitreous to prolapse, stretching the Slinky.

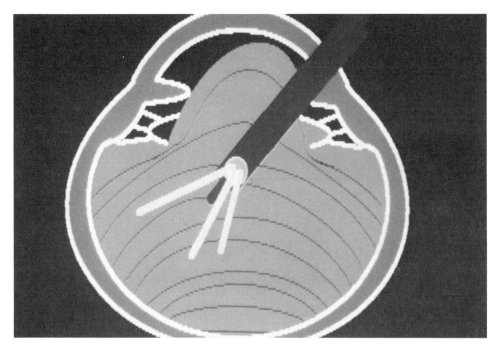

Figure 24–7. Coaxial infusion hydrates the vitreous, forcing more of it into the anterior chamber and stretching the Slinky.

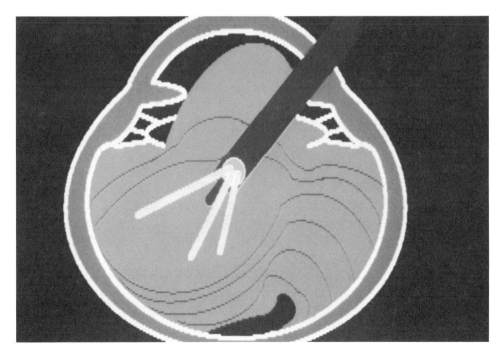

Figure 24–8. Coaxial infusion flushes vitreous toward the anterior chamber, stretching and wiggling the Slinky.

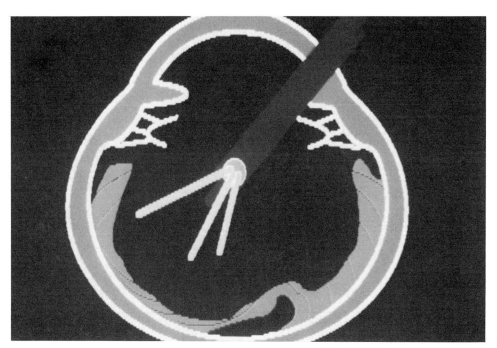

Figure 24–9. After removing all of the vitreous that was washed into the anterior chamber, a lot of the vitreous cavity has been disturbed, including the vitreomacular interface and the vitreous base.

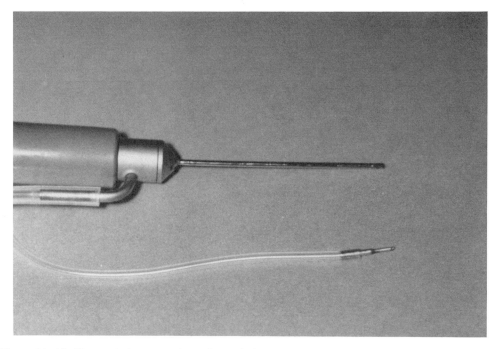

Figure 24–10. Bimanual vitrectomy is performed after separating the infusion from the cutting tip. The separate infusion line is attached to the infusion line of the vitrectomy unit. The aspiration line is still attached to the cutting tip.

Figure 24–11. The infusion is directed parallel to the iris. The cutting tip is laced behind the posterior capsule and the vitreous in the anterior chamber is aspirated downward. The body of the vitreous is not disturbed.

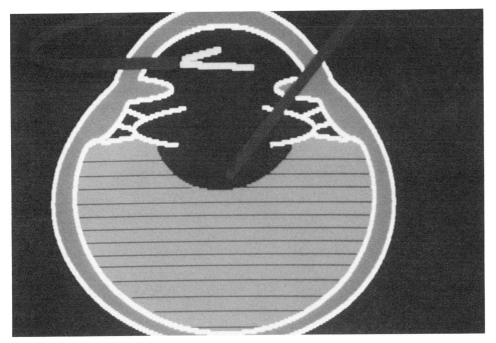

Figure 24–12. At the end of the vitrectomy the vitreous has been removed from the anterior chamber, the capsular bag, and the region just behind the posterior capsule. The rest of the vitreous has not been touched.

Each of these three forces adversely affects the integrity of the vitreous because each of them causes the Slinky to stretch and forces at the Slinky's base to be exerted. It is not surprising that vitrectomy following vitreous loss in cataract surgery has a postoperative complication rate of 30% to 50%.

The best strategy in performing a vitrectomy is to touch the vitreous as little as possible. Remember that phacoemulsification is performed in a closed system, so there is a limit to how much the vitreous can move on its own. The key thought is how to avoid violating any more of the vitreous. If you are able to remove the vitreous from the anterior chamber without disturbing the rest of the vitreous, especially that overlying the vitreous base of the vitreo-macular interface, you should have very few postoperative problems.

The Second Principle of Vitrectomy:

Use a Bimanual Technique With a Separate Infusion Line

To avoid the problems caused by coaxial infusion, remove the coaxial sleeve and replace it with a separate infusion line. A chamber maintainer is a short piece of silicone tubing with a female Leur-lock connector on one end and a short hub on the other. Connect·the Leur-lock connector to the infusion line from the vitrectomy infusion bottle. The tubing can be held in the left hand and the hub placed into the sideport incision when needed (Figure 24–10).

The vitrectomy tip is held in the right hand and passed into the eye, down through the vitreous in the anterior chamber, through the opening in the posterior capsule, and held a millimeter or two behind the posterior capsule. The aspiration port is directed upward toward the cornea.

The strategy is to draw the vitreous in the anterior chamber down to the vitrectomy tip until no more vitreous is in the anterior chamber, and then stopping. We specifically do not want to remove any more vitreous from the vitreous cavity other than cleaning up around the posterior capsule. The body of the vitreous should be unviolated, because, in doing so, we will not disturb the Slinky.

The Third Principle of Vitrectomy

It is Unfortunate if the Slinky is Stretched Due to Vitreous Loss, but it is Worse if the Surgeon Stretches it Any More.

After the vitrectomy tip is in position just behind the posterior capsule, gentle cutting and aspiration can be activated to draw the vitreous down from the anterior chamber to the vitrectomy tip. This does not disturb the rest of the vitreous, but it does soften the eye. After awhile, the eye needs to be firmed up again and irrigation is required.

The irrigation line is placed into the sideport incision, and the flow of the fluid is directed across the anterior chamber in the plane of the iris (Figure 24–11). The goal is to refill the anterior chamber without pushing any of the irrigation fluid behind the posterior capsule. Some admixing of the fluid and the vitreous in the anterior chamber may occur, but that is not important. It will all be removed anyway. The important thing is not to mix irrigation with the vitreous behind the posterior capsule, because that hydration will lead to new vitreous prolapse.

The Fourth Principle of Vitrectomy

The Infusion Maintains the Anterior Chamber. It has No Role in the Vitreous Cavity.

When the vitrectomy has proceeded to the point where the vitreous is out of the anterior chamber, a little more has to be done by the posterior chamber to eliminate any traction or adhesions. Then you are finished. The posterior capsule is usually unaffected by the vitrectomy, and insertion of a posterior chamber intraocular lens can be performed without difficulty, though usually in the ciliary sulcus rather than in the capsular bag. But this depends on the size of the opening in the posterior capsule (Figure 24–12).

If residual cortex is in the capsular bag, it can be removed with the vitrectomy tip using a combination of aspiration and cutting. Because aspiration of the cortex sometimes leads to new vitreous prolapse, the cutter will have to be activated at the first sign of aspiration blockage to sever the vitreous before any tensions are exerted.

Another method of cortical aspiration is to hold the vitreous back with a viscoelastic and remove the cortex with a manual aspirator, such as a cortex cannula on a syringe. If the anterior chamber shallows during cortical aspiration, it will have to be refilled with viscoelastic to maintain a positive pressure against the vitreous.

Results

I reviewed our results in 2,000 consecutive cases of phacoemulsification with lens implantation. The posterior capsule was inadvertently opened in 60 cases (3%), but vitreous loss occurred in only 16 cases (0.8%). I attribute the fact that only 26.6% of cases of open posterior capsules had vitreous loss to the unique advantages of the closed system in phacoemulsification and to the fact that aspiration was terminated before vitreous was aspirated.

All 16 patients with vitreous loss were treated exactly as described in this chapter, and each of the patients had what I considered to be a normal postoperative course. Each of the patients had a corrected visual acuity of 20/20 by the third week postoperatively, and each of the patients were followed for more than a year, during which time each of them maintained 20/20 vision (Table 24–1).

This does not mean that vision loss after vitreous loss during phacoemulsification will not occur, but the incidence of vision loss due to cystoid macular edema is certainly reduced from the usually reported 30% to 50%.

TABLE 24–1 Vitreous Loss during Phacoemulsification

VITREOUS LOSS

NUMBER OF CASES	2,000	100.0%
Posterior Capsule Openings	60	3.0%
Vitreous Loss	16	0.8%
Posterior capsule openings resulting in vitreous loss		26.6%

TIMING OF VITREOUS LOSS

NUMBER OF CASES	16	100.0%
At End Of Cortical Aspiration	4	25.0%
IOL In Capsular Bag	2	
IOL In Sulcus		
During Emulsification	2	12.5%
At End Of Emulsification	10	62.5%

Conclusion

Vitreous loss during phacoemulsification can be limited because the operation is performed in a closed system. If the vitrectomy is also performed in a closed system respecting the integrity of the vitreous body, visual complications can be limited.

The vitrectomy should be performed using a bimanual technique. The main body of the vitreous should be not disturbed, and the vitreous in the anterior chamber should be aspirated downward below the plane of the posterior capsule. Irrigation should be gentle and limited to the anterior chamber.

Using the Slinky for visualization of these principles, we can see that coaxial infusion vitrectomy leads to additional retinal problems because of extension of the capsular tear, hydration of the vitreous, and flushing of the vitreous.

Bimanual vitrectomy is a gentler and safer way to perform vitrectomy after vitreous loss during phacoemulsification because it does not "stretch the Slinky." By following the principles presented, it is possible to have the same morbidity and visual acuity in eyes with vitreous loss as in those without vitreous loss.*

*Just prior to publication of this book, these statistics were updated. In 3,300 cases, there were 32 cases of vitreous loss (1%). Two of the 32 cases developed CME, for a rate of just 6%.

25

Posterior Chamber IOL in Eyes with Large Capsular Tears

Alan W. Solway, MD

When faced with a tear on the posterior capsule, most surgeons would still prefer to implant a posterior chamber IOL. Anterior chamber lenses are often selected, in spite of the potential for long–term complications, because they are relatively easy to insert. Most of the time, one can tell by inspection if there will be enough capsule to support the haptics of a PC lens. But several potential problems are associated with their insertion. One is that residual capsule will tear even further while the lens is being positioned. Another is that the lens haptics or optic will become subluxated into the vitreous. Third is that once the lens is positioned, capsular support will still be inadequate, and more damage will be done when the lens is removed.

These problems can be overcome by using Sheets glides to facilitate lens placement. The glides work in two ways. They act as a scaffold to support the lens optic and guide the haptics into the proper plane. They also serve to protect the residual capsule and keep the vitreous back.

Every effort should be made in all cases to preserve as much capsule as possible. Residual cortex is aspirated manually or with minimal flow and should be aspirated toward the tear rather than away from it. An anterior vitrectomy may be performed with irrigation from a separate port to avoid direct pressure on the tear. If a capsulorhexis is used, it can provide valuable additional anterior capsular support.

If the defect in the posterior capsule is a round hole, the lens can be placed in the capsular bag. In this case, two glides are placed underneath the anterior capsule at 6 o'clock, and the lens is placed between them in a glide-lens-glide sandwich. The top glide guides the lens into the capsular bag, and the bottom glide keeps the vitreous back and protects the posterior capsule.

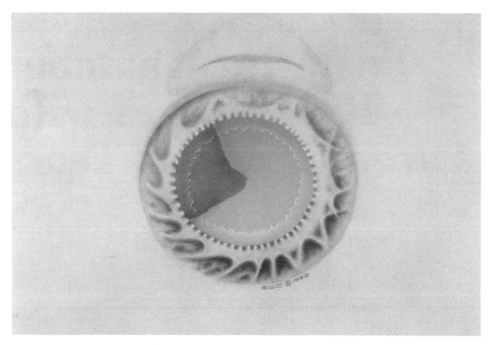

Figure 25–1. A small amount of viscoelastic is used to define the space between the iris and anterior capsule.

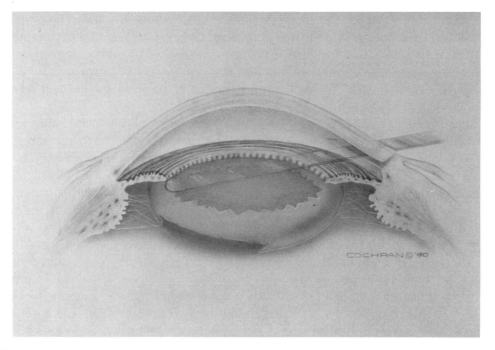

Figure 25–2, 25–3. A Sheets glide is placed behind the iris and in front of the anterior capsule at 6 o'clock.

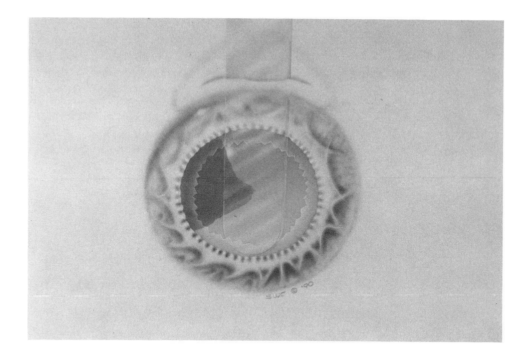

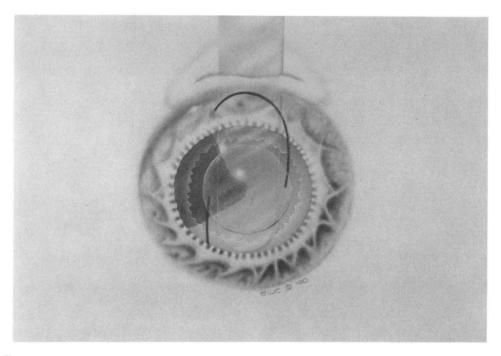

Figure 25-4. The lens is placed in the eye with the inferior haptic between the iris and the glide.

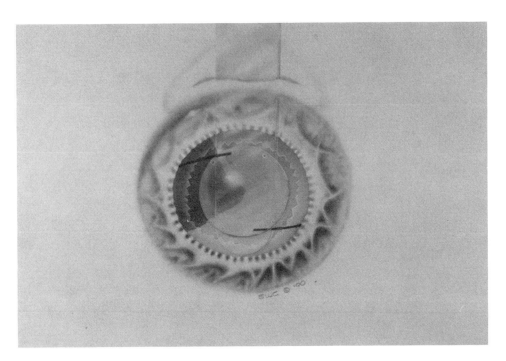

Figure 25–5. The superior haptic is placed with the forceps or dialed into position.

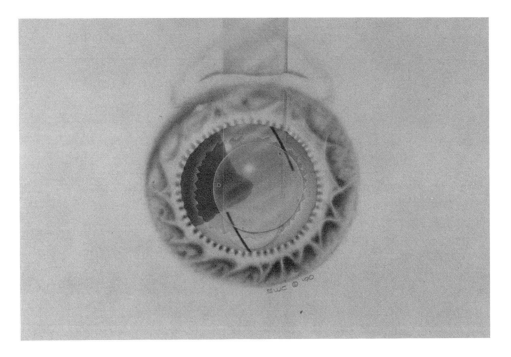

Figure 25–6. With the glide in position, the lens is dialed into a position of maximum capsular support.

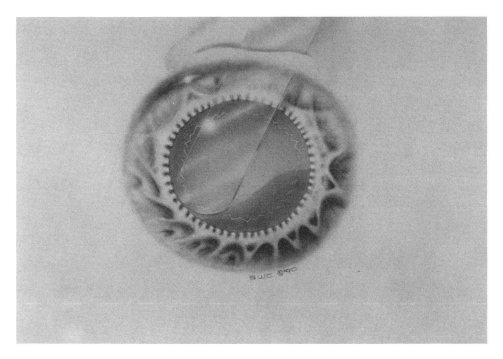

Figure 25–7. If the tear involves the 6 o'clock position, the glide can be placed eccentrically.

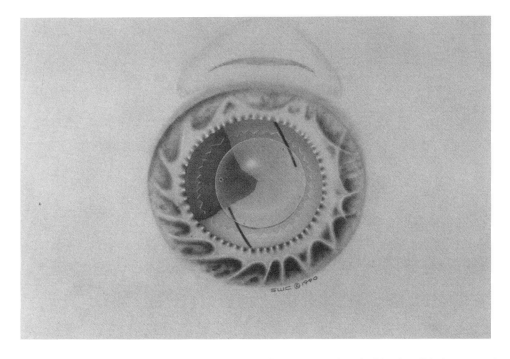

Figure 25–8. After the position of the haptics is verified by retracting the iris, the glide is removed.

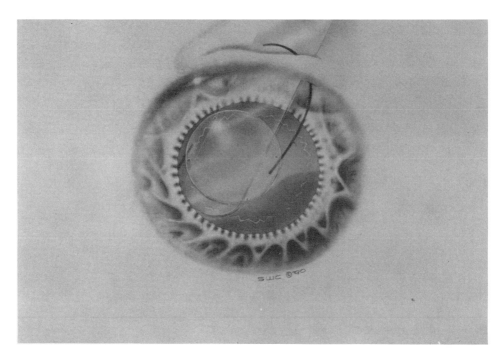

Figure 25-9. The inferior haptic is placed between the iris and the glide.

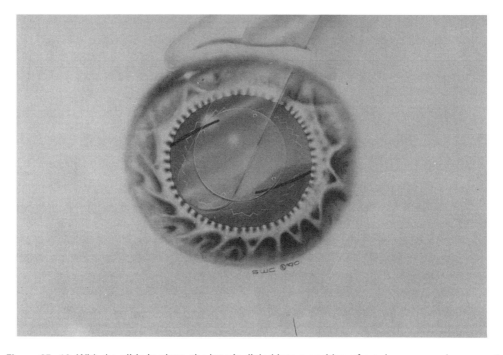

Figure 25-10. With the glide in place, the lens is dialed into a position of maximum capsular support

But, if there is a tear in the posterior capsule, I always place the lens in the ciliary sulcus. I prefer the inherent risk of sulcus fixation to the potential risk of inadequate support or late decentration.

The first step is to use a small amount of viscoelastic to open the space between the anterior capsule and the iris (Figure 25–1). As little as possible is used because it can be difficult to remove at the end of the case. A single Sheets glide is placed between the iris and the anterior capsule at 6 o'clock (Figures 25–2, 25–3). The lens is placed into the eye with the inferior haptic behind the iris and in front of the glide (Figure 25–4). A lens with Prolene haptics is illustrated because the loops are flexible and easy to manipulate.

The superior haptic is either dialed into position or placed with forceps (Figure 25–5). At this point, both haptics are behind the iris and the glide is supporting the lens optic and keeping the vitreous back. Most importantly, the lens haptics are guided into the ciliary sulcus and away from the tear in the capsule.

At this point, with the Sheets glide in place, the lens is dialed into a position where the haptics will receive maximum capsular support (Figure 25–6). Since the glide is coming out of the eye at 12 o'clock, the lens can be rotated only 160 degrees, and the haptics cannot be oriented exactly at 6 o'clock and 12 o'clock. If this orientation is necessary, the glide must be placed eccentrically (Figure 25–7).

Once the haptics are in a position of maximum capsular support, the glide remains in position while the iris is retracted and the position of the haptics is verified. If there is inadequate support, the lens can be removed at this point without damaging the eye. Only when the lens is found to have solid capsular support is the glide finally removed (Figure 25–8).

The technique as illustrated in Figures 25–1 through 25–7 works well for tears at the 12, 9, and 3 o'clock positions. If there is a tear at 6 o'clock or poor anterior capsular delineation, the glide can be placed eccentrically (Figure 25–8). The lens is then placed in front of the glide (Figure 25–9) and rotated to a position of maximum capsular support (Figure 25–10).

Index

Page numbers in italic indicate figures.